# 物理专业英语

## English in Physics

主　编　李淑侠　刘盛春
参　编　鲍东星　孙普男

哈爾濱工業大學出版社

## 内　容　提　要

本书为高等院校物理及相关专业英语基础教材，也适用于从事物理方面理论研究的读者参考。本书共设12个单元，主要内容涉及：力学、热学、光学、电磁学、近代物理等物理学中的五大部分等具有代表性的内容以及一些经典的科普阅读材料。另外，本书的词汇表中列出了基本专业术语的英文解释，便于读者了解和查阅。

**图书在版编目(CIP)数据**

物理专业英语/李淑侠主编. —哈尔滨：哈尔滨工业大学出版社，2005.2(2023.7重印)

ISBN 978-7-5603-2063-2

Ⅰ.物…　Ⅱ.李…　Ⅲ.物理-英语-高等学校-教材　Ⅳ.H31

中国版本图书馆CIP数据核字(2005)第013685号

责任编辑　杨　桦
封面设计　卞秉利
出版发行　哈尔滨工业大学出版社
社　　址　哈尔滨市南岗区复华四道街10号　邮编150006
传　　真　0451-86414749
网　　址　http://hitpress.hit.edu.cn
印　　刷　肇东市一兴印刷有限公司
开　　本　880mm×1230mm　1/32　印张9.375　字数240千字
版　　次　2005年2月第1版　2023年7月第11次印刷
书　　号　ISBN 978-7-5603-2063-2
定　　价　18.00元

# 前　　言

物理专业英语是为提高物理学专业学生专业英文文献的读写能力而开设的必修课，是大学英语学习的重要环节。通过本课程的学习，可以进一步增加学生对专业英语词汇、语法和结构的了解，为拓宽专业知识面并从事专业方面英文资料的查阅、翻译和写作打下坚实的基础。

随着物理学相关学科的理论研究及应用技术的发展，各类专业信息的交流日益广泛，这就要求物理专业人员应具备英文文献熟练获取和共享信息资源的能力。

因而，既具有专业特点，又兼顾通用性的专业英语教材就成为必需。本书正是适合这一需要，经黑龙江大学教师总结多年教学经验而编写的。本书具有以下特点：

1.有较强的系统性和完整性，内容编排由浅入深，符合教学规律，且涵盖了物理学科各个部分，使学习物理学知识和掌握专业英语融为一体。

2.信息涵盖面大，既包括物理学基本知识、基本理论，也涉及相关的课外阅读材料和专业英语词汇。

3.突出了适用性和灵活性，学生可结合自身实际，有针对性地选择学习内容。

4.附有单词表，以方便查阅和对照。

本书内容广泛，知识介绍循序渐进，涵盖了物理学的多个领域，既可作为高等院校物理学各专业本、专科学生的专业英语教材,也可作为物理学领域广大研究人员英语学习的参考书。

书中部分材料参考了一些国外文献,并由作者进行了修改和整理,因篇幅有限,无法一一注释说明,在此一并向原文作者表示谢意!

本书是为解决教学之急需编写的,限于编者水平,尽管做了最大努力,但书中仍难免有疏漏和其他不妥之处,恳请读者提出宝贵意见,以便完善。

**编 者**

**2005 年 1 月**

# CONTENTS

# PART 1
# THE PHYSICAL FUNDAMENTALS of MECHANICS

# 1

## Kinematics

### MECHANICAL PROLOGUE

Mechanics is the study of the motion of material bodies. Historically, it was one of the earliest exact sciences to be developed. Some mechanical principles were known to Greek scientists in the third century B.C.. The tremendous growth of physics since the 1600's began with the discovery of the laws of mechanics by Galileo and Newton. Early successes were in predicting the motions of the moon, the earth, the planets and their satellites (celestial mechanics).

Now we apply some principles to the motions of artificial satellites such as an orbiting Space Shuttle. In general, the principles of mechanics can be applied to

(a) the motions of celestial objects so as to accurately predict events, in some cases many years before they happen, for example, the return of Halley's comet;

(b) the motions of ordinary objects on Earth, for example, an automobile or a thrown baseball;

(c) the behavior of atoms, atomic particles, and subatomic particles, with considerable success.

The term classical mechanics is generally used to differentiate these principles from those newer physical theories, such as relativistic mechanics and quantum mechanics.

Mechanics greatly influenced the growth of later sciences such as sound and electricity. It may be said that mechanics furnishes the basic concepts of the whole physics, so quite naturally, the study of physics begins here.

## 1.1 MECHANICAL MOTION

Mechanical motion is the simplest form motion of matter. It consists in the movement of bodies or their parts relative to one another. We can see movements of bodies everywhere in our ordinary life. This is why mechanical notions are so clear. This also explains the fact that mechanics was the first of all the natural sciences to be developed very broadly.

A combination of bodies separated for consideration is called a mechanical system. The bodies to be included in a system depend on the nature of the problem being solved. In a particular case, a system may consist of a single body.

It was indicated above that motion in mechanics is defined as the change in the mutual arrangement of the bodies. If we imagine a separate isolated body in a space where no other bodies are present, then we cannot speak of the motion of the body because there is nothing with respect to which the body could change its position. It thus follows that if we intend to study the motion of a body, then we must indicate with respect to what other bodies the given motion occurs.

Motion occurs both in space and in time (space and time are inalienable forms of existence of matter). Consequently, to describe motion, we must also determine time. We use a timepiece (watch or clock) for this purpose.

A combination of bodies that are stationary relative to one another with respect to which motion is being considered and a timepiece indicating the time forms a reference frame.

The motion of the same body relative to different reference frames may have a different nature. For example, let us imagine a train gaining speed. Suppose that a passenger is walking with a constant velocity along the corridor of one of the cars of the train. The motion of the passenger relative to the car will be uniform, and relative to the Earth's surface it will be accelerated.

To describe the motion of a body means to indicate for every moment of time the position of the body in space and its velocity. To set the state of a mechanical system, we must indicate the positions and the velocities of all the bodies forming the system. A typical problem of mechanics consists in determining the states of a system at all the following moments of time $t$ when we know the state of the system at a certain initial moment to and also the laws governing the motion.

It must be noted that no physical problem can be solved absolutely exactly. An approximate solution is always obtained. The degree of approximation is determined by the nature of the problem and the object to be achieved. In solving a problem approximately, we disregard the factors that are not significant in the given case. For example, we may often disregard the dimensions of the body whose motion is being studied. For instance, it is quite possible to disregard the earth's dimensions when treating its motion about the sun. This allows us to considerably simplify our description of the motion because the earth's position in space can be determined by a single point.

A body whose dimension may be disregarded in the conditions of a given problem is called a **point particle** (or simply a **particle**). Whether or not we may consider a given body as a particle depends not on the dimensions of the body, but on the conditions of the problem. The same body in some cases may be treated as a particle, but in others it must be considered as a extended body.

When speaking about a body as a particle, we disengage ourselves from its dimensions. Another abstraction which we have to do with in mechanics is a perfectly rigid body. Absolutely undeformable bodies do not exist in nature. Any body deforms to a greater or smaller extent, i.e. changes its shape and dimensions, under the action of forces applied to it. The deformations of bodies when considering their movements may often be disregarded, however. If this is done, the body is called perfectly rigid. Thus, a body whose deformations may be disregarded in the conditions of a given problem is called a perfectly rigid, or simply a rigid body.

Any motion of a rigid body can be resolved into two basic kinds of motion translational motion and circular motion.

Translational motion(translation) is defined as motion in which any straight line associated with the moving body remains parallel to itself (Fig.1.1).

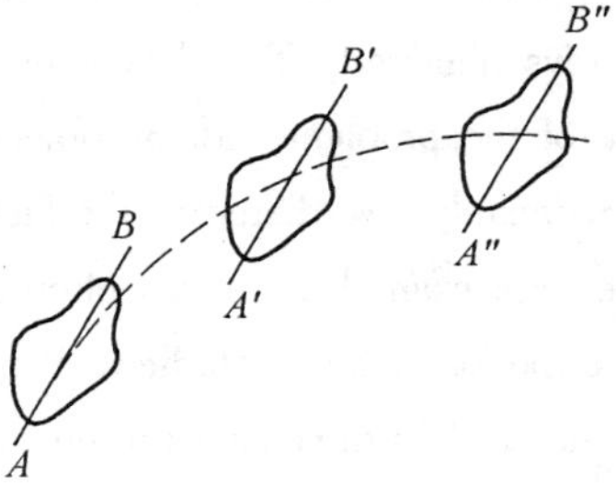


**Fig.1.1**

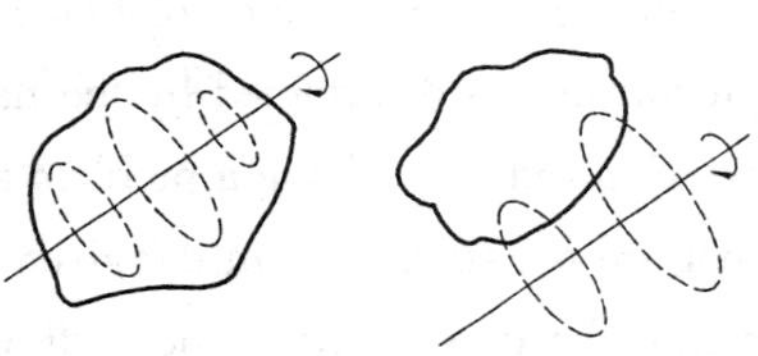

(a) (b)

**Fig.1.2**

In circular motion(rotation), all the points of a body move in circles whose centers are on a single straight line called the axis of rotation(Fig. 1.2). The axis of rotation can be outside a body(Fig. 1.2).

Since when treating a body is a particle we ignore its length, the concept of circular motion about an axis passing through such a body can not be applied to it.

To acquire the possibility of describing motion quantitatively, we have to associate a coordinate system (for example a Cartesian one) with the bodies forming a reference frame. Hence, the position of a particle can be determined by setting the three numbers $x$, $y$, and $z$—the Cartesian coordinates of the particle. A coordinates system can be made by forming a rectangular lattice from identical rods or rules graduated to a definite scale (Fig. 1.3). Identical clocks system can be made by forming a rectangular lattice form identical rods or rules graduated to a definite scale (Fig. 1.3). Identical clocks synchronized with one another must be placed at the lattice points. The position of a particle and the moment of time corresponding to this position are recorded on the graduated rods and the clock closest to the particle.

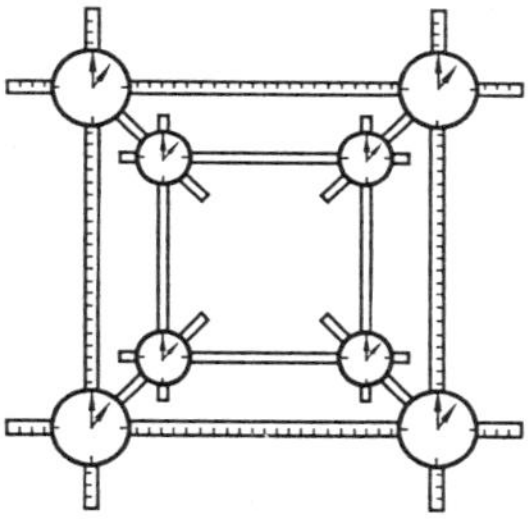

**Fig. 1.3**

It is simpler to treat a point particle than an extended body. We shall therefore first study the mechanics of a particle, and then go over to the mechanics of a rigid body. We shall start with kinematics, and then delve into dynamics. We remind our reader that kinematics studies the motion of bodies without regard to what causes this motion. Dynamics studies the motion of bodies with a view to what causes this motion to have the nature it does, i.e. with a view to the interactions between bodies.

## 1.2 VECTORS

Vectors are defined as quantities characterized by a numerical value and a direction and, also, as ones that are added according to the triangle or parallelogram method. The last requirement is a very significant one. We can indicate quantities characterized by a numerical value and a sense of direction, but that are added in a different way than vectors. We shall take as an example the rotation of a body about an axis through the finite angle $\varphi$. Such rotation can be depicted in the form of a segment of length $\varphi$ directed along the axis about which rotation is occurring and pointing in a direction associated with what of rotation according to the right-hand screw rule. The top portion of Fig.1.4 shows two consecutive turns of the sphere through the angles $\pi/2$ depicted by the segments $\varphi_1$ and $\varphi_2$. The first turn about axis 1 – 1 transfers point $A$ of the sphere to position $A'$, and the second turn about axis 2 – 2 transfers it to the position $A''$. The same result, i.e transfer of point $A$ to position $A''$, can be achieved by turning the sphere about axis 3 – 3 (see the bottom portion of Fig.1.4) through the angle $\pi$. Hence, such a turn should be considered as the sum of the turns $\varphi_1$ and $\varphi_2$. It cannot be obtained from the segments and

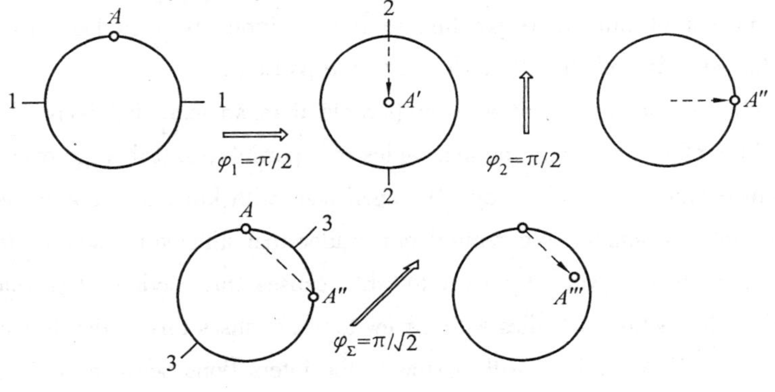


**Fig.1.4**

however, by adding them according to the parallelogram method. Such addition gives a segment of length $\pi/\sqrt{2}$ instead of the required length $\pi$. Rotation through the angle transfers point $A$ to point $A'''$. It thus follows that the turns through finite angles depicted by the directed segments do not have the properties of vectors.

The numerical value of a vectors is called its magnitude. Figuratively speaking, the magnitude of a vector indicates its length. The magnitude of a vector is a scalar, and always a positive one.

Vectors are represented graphically by arrows. The length of an arrow determines to the established scale the magnitude of the relevant vector, and the arrow points in the direction of the vector.

Vectors are customarily distinguished by setting their symbols in boldface type, for example, $\boldsymbol{a}$, $\boldsymbol{b}$, $\boldsymbol{v}$ and $\boldsymbol{F}$. The same symbols set in italics signify the magnitude of the relevant vectors, for example, $a$ is the magnitude of the vector $\boldsymbol{a}$. It is sometimes necessary to express the magnitude by placing a vertical bar (an absolute value sign) on each side of the symbol for the vector. Thus, $|\boldsymbol{a}|$ is the magnitude of the vector $\boldsymbol{a}$. This representation is used, for example, to show the magnitude of the sum of the vectors $\boldsymbol{a}_1$ and $\boldsymbol{a}_2$:

$$|\boldsymbol{a}_1 + \boldsymbol{a}_2| = \text{magnitude of the vector } (\boldsymbol{a}_1 + \boldsymbol{a}_2) \tag{1.1}$$

In this case, the notation $\boldsymbol{a}_1 + \boldsymbol{a}_2$ signifies the sum of the magnitudes of the vectors being added, which in general does not equal the magnitude of the sum of vectors (the two sums will be equal only when the vectors being added have the same direction).

Vectors directed along parallel straight lines (in the same or in opposite directions) are called collinear. Vectors in parallel planes are called coplanar.

Collinear vectors can be arranged along the same straight line and coplanar vectors can be brought into one plane by parallel translation. Collinear vectors equal in magnitude and having the same direction are considered to equal each other.

Vectors addition and subtraction. It is more convenient to add vectors in practice without constructing a parallelogram. Examination of Fig. 1.5 shows that we can achieve the same result if we bring the tail of the second vector in contact with the tip of the first one, and then draw the resultant vector from the tail of the first vector to the tip of the second one. It is very good to use this procedure when we have to add more than two vectors (Fig.1.6).

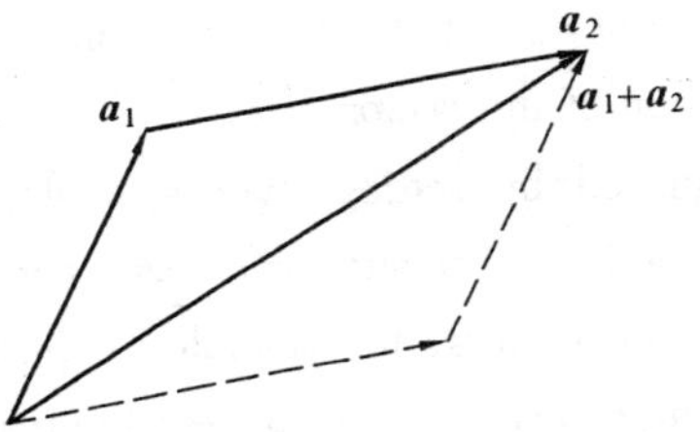


Fig.1.5

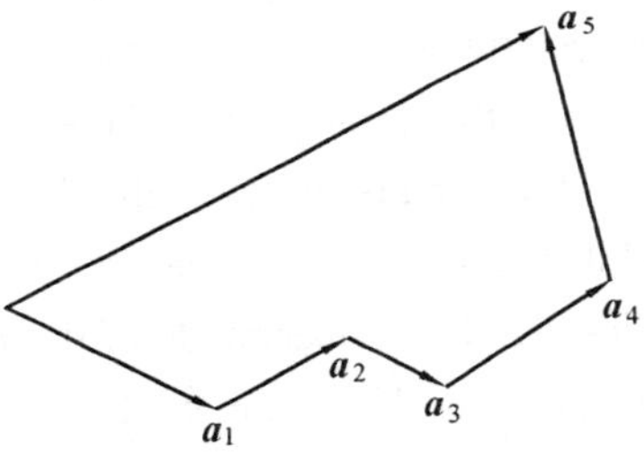


Fig.1.6

The magnitude of the difference of two vector, like the magnitude of a sum [see Eq.1.1] may be written only with the aid of vertical bars:

$$|\boldsymbol{a}_1 - \boldsymbol{a}_2| = \text{magnitude of the vector } (\boldsymbol{a}_1 - \boldsymbol{a}_2) \qquad (1.2)$$

Because the notation $a_1 - a_2$ signifies the difference of magnitudes of the vectors $\boldsymbol{a}_1$ and $\boldsymbol{a}_2$, which generally speaking does not equal the magnitude of the vector difference.

## 1.3 VELOCITY AND SPEED

A point particle in motion travels along a certain line. The latter is called its path or trajectory. Depending on the shape of a trajectory, we distinguish rectilinear or straight motion, circular motion, curvilinear motion, etc.

Assume that a point particle (in the following we shall call it simply a particle for brevity's sake) traveled along a certain trajectory from point 1 to 2 (Fig.1.7). The path between points 1 and 2 measured along the trajectory is called the distance traveled by the particle. We shall denote it by the symbol $s$.

The straight line between points 1 and 2, i.e. the shortest distance between these points, is called the displacement of the particle. We shall denote it by the symbol $\boldsymbol{r}_{12}$. Let us assume that a particle completes two successive displacements $\boldsymbol{r}_{12}$ and $\boldsymbol{r}_{13}$(Fig.1.8). It is natural to call such a displacement $\boldsymbol{r}_{13}$ the sum of the first two that lead to the same result as they do together. Thus, displacement are characterized by magnitude and direction and, besides, are added by using the parallelogram method. Hence, it follows that displacement is a vector.

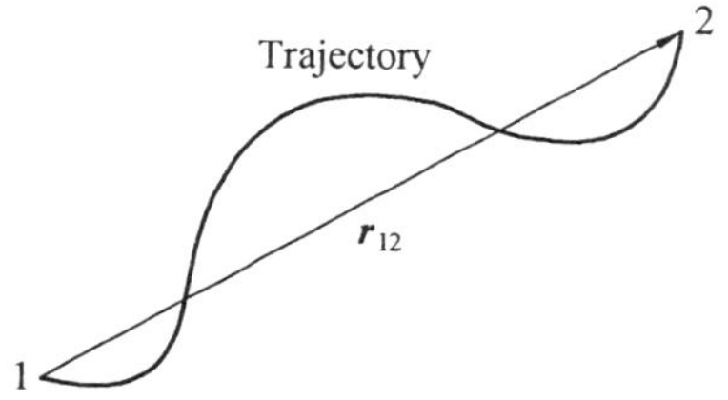


**Fig.1.7**

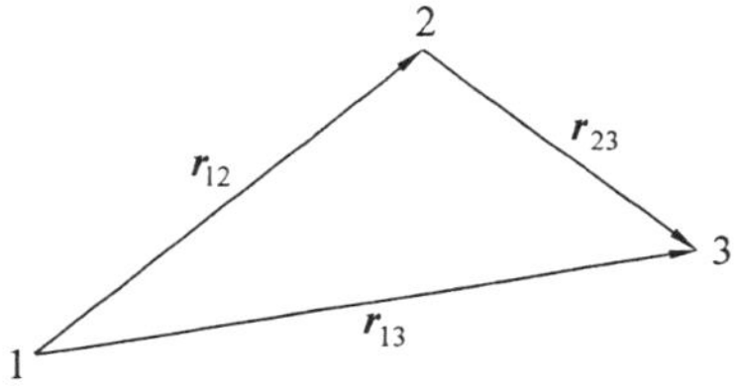


**Fig.1.8**

In everyday life, we use the terms speed and velocity interchangeably, but in physics there is an important distinction between them. Speed depends on the distance travelled, and velocity on the displacement. Speed is the distance travelled by a particle in unit time. If a particle travels identical distances during equal time intervals that may be as small as desired, its motion is called uniform. In this case, the speed of the particle at each moment can be calculated by dividing the distance $s$ by the time $t$.

Velocity is a vector quantity characterizing not only how fast a particle travels along its trajectory, but also the direction in which the particle moves at each moment. Let us divide a trajectory into infinitely small portions of length d$s$. An infinitely small displacement d$\boldsymbol{r}$ corresponds to each of these portions (Fig. 1.9). Dividing this displacement by the corresponding time interval d$t$, we get the instantaneous velocity at the given point of the trajectory:

$$\boldsymbol{v} = \frac{\mathrm{d}\boldsymbol{r}}{\mathrm{d}t} = \dot{\boldsymbol{r}} \tag{1.3}$$

Thus, the velocity is the derivative of the position vector of the particle with respect to time. The displacement d$\boldsymbol{r}$ coincides with an infinitely small element of the trajectory. Consequently, the vector $\boldsymbol{v}$ is directed along a tangent to the trajectory (Fig. 1.9).

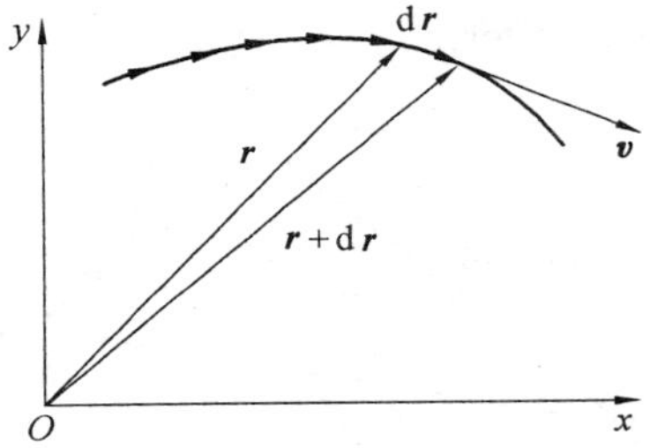


**Fig. 1.9**

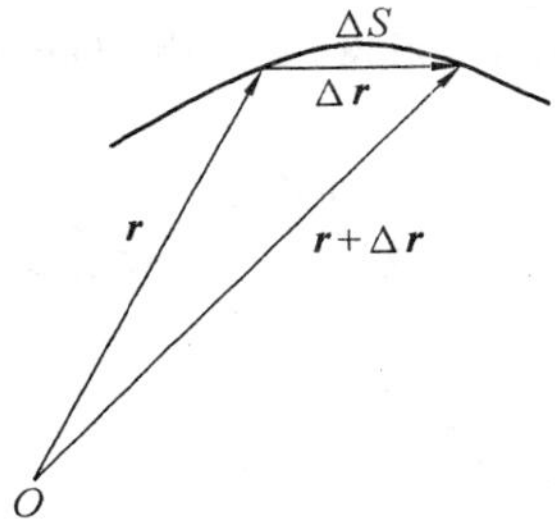


**Fig. 1.10**

Reasoning more strictly, to derive formula (1.3) we must proceed as follows. Having fixed a certain moment of time $t$, let us consider the increment of the position vector $\Delta\boldsymbol{r}$ during the small time interval $\Delta t$ following $t$ (Fig. 1.10). The ratio $\Delta\boldsymbol{r}/\Delta t$ gives the average value of the velocity during the time $\Delta t$. If we take smaller and smaller intervals $\Delta t$, the ratio $\Delta\boldsymbol{r}/\Delta t$ in the limit will give us the value of the velocity $\boldsymbol{v}$ at the moment $t$:

$$\boldsymbol{v} = \lim_{\Delta t \to 0} \frac{\Delta\boldsymbol{r}}{\Delta t} = \frac{\mathrm{d}\boldsymbol{r}}{\mathrm{d}t} \tag{1.4}$$

We have arrived at formula (1.3).

Let us find the magnitude of the expression (1.4), i.e. the magnitude of the velocity $\boldsymbol{v}$:

$$v = |\boldsymbol{v}| = \left| \lim_{\Delta t \to 0} \frac{\Delta \boldsymbol{r}}{\Delta t} \right| = \lim_{\Delta t \to 0} \frac{|\Delta \boldsymbol{r}|}{\Delta t} \tag{1.5}$$

We cannot write $\Delta r$ instead of $|\Delta \boldsymbol{r}|$ in this formula. The vector $\Delta \boldsymbol{r}$ is in essence the difference between two vectors ($\boldsymbol{r}$ at the moment $t + \Delta t$ minus $\boldsymbol{r}$ at the moment $t$). Therefore, its magnitude may be written only with the aid of vertical bars [see Eq. 1.2]. The symbol $|\Delta \boldsymbol{r}|$ signifies the magnitude of the increment of the vector $\boldsymbol{r}$, whereas $\Delta r$ is the increment of the magnitude of the vector $\boldsymbol{r}$, i.e. $\Delta |\boldsymbol{r}|$. These two quantities, generally speaking, do not equal each other:

$$|\Delta \boldsymbol{r}| \neq \Delta |\boldsymbol{r}| = \Delta r$$

The following example will illustrate this. Assume that the vector $\boldsymbol{r}$ receives such an increment $\Delta \boldsymbol{r}$ that its magnitude does not change i.e. $|\boldsymbol{r} + \Delta \boldsymbol{r}| = |\boldsymbol{r}|$ (Fig. 1. 11). Consequently, the increment of the magnitude of the vector equals zero ($\Delta |\boldsymbol{r}| = \Delta r = 0$). At the same time, the magnitude of the increment of the vector $\boldsymbol{r}$, i.e. $|\Delta \boldsymbol{r}|$, differs from zero (it equals the length of 2 – 3). What has been said above holds for any vector $\boldsymbol{a}$: in the general case $|\Delta \boldsymbol{a}| \neq \Delta a$.

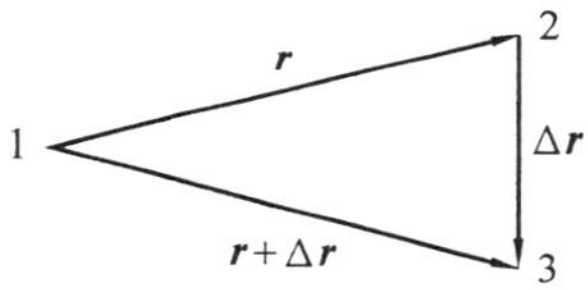

**Fig.1.11**

Inspection of Fig. 1.10 shows that the distance $\Delta s$, generally speaking, differs in value from the magnitude of the displacement $|\Delta \boldsymbol{r}|$. If we take increments of the distance $\Delta s$ and the displacement $\Delta \boldsymbol{r}$ corresponding to smaller and smaller time intervals $\Delta t$, then the difference between $\Delta s$ and $|\Delta \boldsymbol{r}|$ will diminish, and their ratio in the limit will become equal to unity:

$$\lim_{\Delta t \to 0} \frac{\Delta s}{|\Delta \boldsymbol{r}|} = 1$$

On these grounds, we can substitute $\Delta s$ for $|\Delta \boldsymbol{r}|$ in formula (1.5), which gives us the expression

$$v = \lim_{\Delta t \to 0} \frac{\Delta s}{\Delta t} = \frac{\mathrm{d}s}{\mathrm{d}t} \tag{1.6}$$

Thus, the magnitude of the velocity equals the derivative of the distance with respect to time.

It is evident that the quantity which in everyday life we call the speed is actually the magnitude of the velocity $\boldsymbol{v}$. In uniform motion, the magnitude of the velocity remains constant ($v$ = const), whereas the direction of the vector $\boldsymbol{v}$ changes arbitrarily (in particular it may be constant).

In accordance with Eq. (1.3), the elementary displacement of a particle is

$$\mathrm{d}\boldsymbol{r} = \boldsymbol{v}\mathrm{d}t \tag{1.7}$$

Sometimes for clarity's sake, we shall denote an elementary displacement by the symbol $\mathrm{d}\boldsymbol{s}$, i.e. write Eq. (1.7) in the form

$$\mathrm{d}\boldsymbol{s} = \boldsymbol{v}\mathrm{d}t \tag{1.8}$$

The velocity vector, like any other vector, can be represented in the form

$$\boldsymbol{v} = v_x\boldsymbol{e}_x + v_y\boldsymbol{e}_y + v_z\boldsymbol{e}z \tag{1.9}$$

where $v_x$, $v_y$, $v_z$ are the projections of the vector $\boldsymbol{v}$ onto the coordinate axes. At the same time, the vector $\dot{\boldsymbol{r}}$ equal to $\boldsymbol{v}$, can be written as follows:

$$\dot{\boldsymbol{r}} = \dot{x}\boldsymbol{e}_x + \dot{y}\boldsymbol{e}y + \dot{z}\boldsymbol{e}_z \tag{1.10}$$

It follows from a comparison of Eq. (1.9) and (1.10) that

$$v_x = \dot{x}, \quad v_y = \dot{y}, \quad v_z = \dot{z} \tag{1.11}$$

Consequently, the projection of the velocity vector onto a coordinate axis equals the time derivative of the relevant coordinate of the moving particle. We get:

$$v = \sqrt{\dot{x}^2 + \dot{y}^2 + \dot{z}^2} \tag{1.12}$$

The velocity vector can be written in the form $\boldsymbol{v} = v\boldsymbol{e}_v$, where $v$ is the magnitude of the velocity, and $\boldsymbol{e}_v$ is the unit vector of $\boldsymbol{v}$. Let us introduce the unit vector $\boldsymbol{\tau}$ of the tangent to a trajectory with its sense the same as that of $\boldsymbol{v}$. Hence, obviously, the unit vectors $\boldsymbol{e}_v$, and $\boldsymbol{\tau}$ will coincide, and we can write the following expression:

$$\boldsymbol{v} = v\boldsymbol{e}_v = v\boldsymbol{\tau} \tag{1.13}$$

Let us obtain still another expression for $\boldsymbol{v}$. For this purpose we shall introduce the position vector in the form of $\boldsymbol{r} = r\boldsymbol{e}_r$ into Eq. (1.3). We have

$$\boldsymbol{v} = \dot{\boldsymbol{r}} = \dot{r}\boldsymbol{e}_r + r\dot{\boldsymbol{e}}_r \tag{1.14}$$

We shall limit ourselves, for simplicity, to the case when the trajectory is a plane curve, i.e. a curve such that all its points are in a single plane. Let this plane be the plane $x, y$. In Eq. (1.14), the vector $\boldsymbol{v}$ is written in the form of two components (Fig. 1.12). The first of them, which we shall designate $\boldsymbol{v}_r$, is

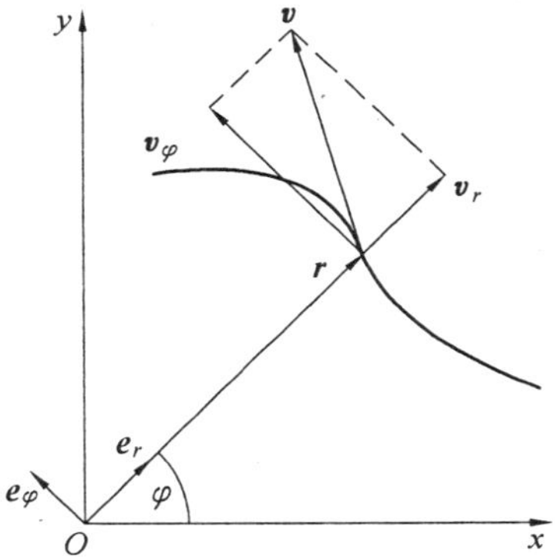


**Fig.1.12**

$$\boldsymbol{v}_r = \dot{r}\boldsymbol{e}_r \tag{1.15}$$

It is directed along the position vector $\boldsymbol{r}$ and characterizes the rate of change of the magnitude of $\boldsymbol{r}$. The second component, which we shall designate $\boldsymbol{v}_\varphi$, is

$$\boldsymbol{v}_\varphi = r\dot{\boldsymbol{e}}_r \tag{1.16}$$

It characterizes the rate of change of the direction of the position vector.

We can write that

$$\dot{\boldsymbol{e}}_r = \frac{\mathrm{d}\varphi}{\mathrm{d}t}\boldsymbol{e}_\varphi = \dot{\varphi}\boldsymbol{e}_\varphi$$

where $\varphi$ is the angle between the position vector and the $x$-axis, and $\boldsymbol{e}_\varphi$ is a unit vector perpendicular to the position vector with its sense in the direction of growth of the angle $\varphi$. Using this value in Eq. (1.16), we get:

$$\boldsymbol{v}_\varphi = r\dot{\varphi}\boldsymbol{e}_\varphi \tag{1.17}$$

We have introduced the symbols $\boldsymbol{v}_\varphi$ and $\boldsymbol{e}_\varphi$ to underline the fact that the component $\boldsymbol{v}_\varphi$ and the corresponding unit vector are related to a change in the angle $\varphi$.

The vectors $\boldsymbol{v}_r$ and $\boldsymbol{v}_\varphi$ are obviously mutually perpendicular. Hence,

$$v = \sqrt{v_r^2 + v_\varphi^2} = \sqrt{\dot{r}^2 + r^2\dot{\varphi}^2} \tag{1.18}$$

Now let us consider how to calculate the distance travelled by a particle from the moment of time $t_1$ to $t_2$ if we know the speed at each moment. Let us divide the interval $t_1 - t_2$ into $N$ small, but not necessarily equal intervals: $\Delta t_1, \Delta t_2, \cdots, \Delta t_N$. The total distances travelled by a particle can be represented as the sum of the distances, $\Delta s_1, \Delta s_2, \cdots, \Delta s_N$ travelled during the relevant time intervals $\Delta t$

$$s = \Delta s_1 + \Delta s_2 + \cdots + \Delta s_N = \sum_{i=1}^{N} \Delta s_i$$

In accordance with formula (1.6), each of the addends can approximately be represented in the form

$$\Delta s_i \approx v_i \Delta t_i$$

where $\Delta t_i$ is the time interval during which the distance $\Delta s_i$ was travelled, and $v_i$ is one of the values of the speed during the time $\Delta t_i$. Hence,

$$s_i \approx \sum_{i=1}^{N} v_i \Delta t_i \tag{1.19}$$

This expression will be obeyed more accurately with diminishing time intervals $\Delta t_i$. In the limit when all the $\Delta t_i$'s tend to zero (the number of intervals $\Delta t_i$ will correspondingly grow unlimitedly), the approximate equation will become accurate:

$$s = \lim_{\Delta t_i \to 0} \sum_{i=1}^{N} v_i \Delta t_i$$

This expression is a definite integral of the function $v(t)$ taken within the limits from $t_1$ to $t_2$. Thus, the distance travelled by a particle during the interval from $t_1$ to $t_2$ is

$$s = \int_{t_1}^{t_2} v(t)\mathrm{d}t \tag{1.20}$$

It must be underlined that here we are speaking of the speed. If we take an integral of the velocity $\boldsymbol{v}(t)$, we get the vector of the displacement of the particle from the point where it was at the moment $t_1$ to the point where it was at the moment $t_2$[see Eq. (1.7)]:

$$\int_{t_1}^{t_2} \boldsymbol{v}(t)\mathrm{d}t = \int_{t_1}^{t_2} \mathrm{d}\boldsymbol{r} = \boldsymbol{r}_{12} \tag{1.21}$$

If we plot the dependence of $v$ on $t$ (Fig. 1.13), then the distance travelled can be represented as the area of the figure confined between the curve $v(t)$, the straight lines $t = t_1$ and $t = t_2$, and the $t$-axis. Indeed, the product $v_i\Delta t_i$ numerically equals the area of the $i$-th strip. The sum (1.19) equals the area of the figure confined on top by the broken line formed by the top edges of all such strips. When all the $\Delta t_i$'s tend to zero, the width of a strip diminishes (their number grows simultaneously), and the broken line will coincide with the curve $v = v(t)$ in the limit. Thus, the distance travelled during the time from the moment $t_1$ to the moment $t_2$ numerically equals the area confined between the curve of the function $v = v(t)$ the time axis, and the straight lines $t = t_1$ and $t = t_2$.

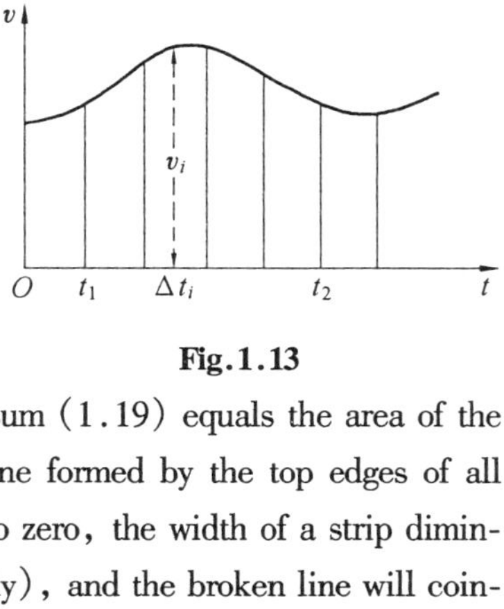

**Fig.1.13**

It should be noted that the average value of the speed during the time from $t_1$ to $t_2$, by definition, is

$$\langle v \rangle = \frac{s}{t_2 - t_1}$$

(The symbol $\langle\rangle$ embracing the $v$ indicates an average). Introducing into this equation the expression (1.20) for $s$, we get

$$\langle v \rangle = \frac{1}{t_1 - t_2}\int_{t_1}^{t_2} v(t)\mathrm{d}t \tag{1.22}$$

The average values of any scalar or vector functions are calculated in a similar way. For example, the average value of the velocity is [see (Eq. 1.21)]

$$\langle \boldsymbol{v} \rangle = \frac{1}{t_2 - t_1}\int_{t_1}^{t_2} \boldsymbol{v}(t)\mathrm{d}t = \frac{\boldsymbol{r}_{12}}{t_2 - t_1} \tag{1.23}$$

The average value of the function $y(x)$ within the interval from $x_1$ to $x_2$ is determined by the expression

$$\langle y \rangle = \frac{1}{x_2 - x_1}\int_{x_1}^{x_2} y(x)\mathrm{d}x \tag{1.24}$$

## 1.4 ACCELERATION

The velocity $\boldsymbol{v}$ of a particle can change with time both in magnitude and in direction. The rate of change of the vector $\boldsymbol{v}$, like the rate of change of any function of time, is determined by the derivative of the vector $\boldsymbol{v}$ with respect to $t$. Denoting this derivative by the symbol $\boldsymbol{a}$, we get:

$$\boldsymbol{a} = \lim_{\Delta t \to 0}\frac{\Delta \boldsymbol{v}}{\Delta t} = \frac{\mathrm{d}\boldsymbol{v}}{\mathrm{d}t} = \dot{\boldsymbol{v}} \tag{1.25}$$

The quantity determined by formula (1.25) is called the acceleration of the particle.

It must be noted that the acceleration $\boldsymbol{a}$ plays the same part with respect to $\boldsymbol{v}$ as the vector $\boldsymbol{v}$ does with respect to the position vector $\boldsymbol{r}$.

Equal vectors have identical projections onto the coordinate axes. Consequently, for example,

$$a_x = \left(\frac{\mathrm{d}v}{\mathrm{d}t}\right)_{\mathrm{pr},x} = \frac{\mathrm{d}v_x}{\mathrm{d}t} = \dot{v}_x$$

At the same time according to Eq. (1.11), we have $v_x = \dot{x} = \mathrm{d}x/\mathrm{d}t$. Therefore,

$$\frac{\mathrm{d}v_x}{\mathrm{d}t} = \frac{\mathrm{d}}{\mathrm{d}t}\left(\frac{\mathrm{d}x}{\mathrm{d}t}\right) = \frac{\mathrm{d}^2x}{\mathrm{d}t^2} = \ddot{x}$$

What we have obtained is that the projection of the acceleration vector onto the $x$-axis equals the second derivative of the coordinate $x$ with respect to time: $a_x = \ddot{x}$. Similar expressions are obtained for the projections of the acceleration onto the $y$-axes and $z$-axes. Thus,

$$a_x = \ddot{x}, \quad a_y = \ddot{y}, \quad a_z = \ddot{z} \tag{1.26}$$

Using Eq. (1.13) for $\boldsymbol{v}$ in Eq. (1.25), we get:

$$\boldsymbol{a} = \frac{d}{\mathrm{d}t}(v\dot{\boldsymbol{\tau}}) \tag{1.27}$$

We remind our reader that $\boldsymbol{\tau}$ is the unit vector of a tangent to a trajectory having the same direction as $\boldsymbol{v}$. According to

$$\frac{\mathrm{d}}{\mathrm{d}t}(\varphi\boldsymbol{a}) = \varphi\frac{\mathrm{d}\boldsymbol{a}}{\mathrm{d}t} + \boldsymbol{a}\frac{\mathrm{d}\varphi}{\mathrm{d}t} = \varphi\dot{\boldsymbol{a}} + \dot{\varphi}\boldsymbol{a}$$

$$\boldsymbol{a} = \dot{v}\boldsymbol{\tau} + v\dot{\boldsymbol{\tau}} \tag{1.28}$$

Hence, the vector $\boldsymbol{a}$ can be represented in the form of the sum of two components. One of them has the direction $\boldsymbol{\tau}$, i.e. is tangent to the trajectory. It is therefore designated $\boldsymbol{a}_\tau$ and is called the tangential acceleration. It equals

$$\boldsymbol{a}_\tau = \dot{v}\boldsymbol{\tau} \tag{1.29}$$

The second component equal to $v\dot{\boldsymbol{\tau}}$ is directed, as we shall show below, along a normal to the trajectory. It is therefore designated $\boldsymbol{a}_n$ and is called the normal acceleration. Thus,

$$\boldsymbol{a}_n = v\dot{\boldsymbol{\tau}} \tag{1.30}$$

In studying the properties of the two components, we shall restrict ourselves for the sake of simplicity to the case when the trajectory is a plane curve.

The magnitude of the tangential acceleration in Eq. (1.29) is

$$a_\tau = |\dot{v}| \tag{1.31}$$

If $\dot{v} > 0$ (the velocity grows in magnitude), then the vector $\boldsymbol{a}_\tau$ has the same direction as $\boldsymbol{\tau}$ (i.e. the same direction as $\boldsymbol{v}$). If $\dot{v} < 0$ (the velocity decreases with time), then the vectors $\boldsymbol{v}$ and $\boldsymbol{a}_\tau$ have opposite directions. In uniform motion, $\dot{v} = 0$, and, therefore, tangential acceleration is absent.

To determine the properties of the normal acceleration [Eq. (1.30)], we must find out what $\dot{\boldsymbol{\tau}}$ the rate change with time of the direction of tangent to the trajectory, is determined by. It is easy to understand that this rate will grow with an increasing curvature of the trajectory and a higher velocity of a particle along it.

The degree of bending of a plane curve is characterized by its curvature $C$ determined by the expression below.

$$C = \lim_{\Delta s \to 0} \frac{\Delta \varphi}{\Delta s} = \frac{d\varphi}{ds} \tag{1.32}$$

where $\Delta\varphi$ is the angle between tangents to the curve at points spaced $\Delta s$ apart (Fig. 1.14). Thus, the curvature determines the rate of turning of a tangent in motion along a curve.

The reciprocal of the curvature $C$ is called the radius of curvature at the given point of the curve and is designated $R$:

$$R = \frac{1}{C} = \lim_{\Delta\varphi \to 0} \frac{\Delta s}{\Delta \varphi} = \frac{ds}{d\varphi} \tag{1.33}$$

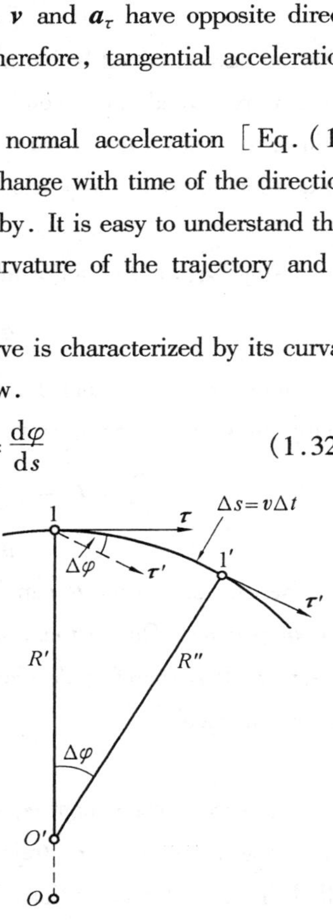

Fig.1.14

The radius of curvature is the radius of a circle that coincides at the given spot with the curve on an infinitely small portion of it. The center of curvature is defined as the center of curvature for the given point of the curve.

The radius and center of curvature at point 1 (Fig. 1.14) can be determined as follows. Take point 1′near point 1. Draw the tangents $\tau$ and $\tau'$ at these points.

The perpendiculars to the tangents will intersect at a certain point $O'$. We must note that for a curve which is not a circle the distances $R'$ and $R''$ will differ somewhat from each other. If point 1′ is brought closer to point 1, the point of intersection $O'$ of the perpendiculars will move along the straight line $R'$ and in the limit will be at point $O$. It is exactly the latter that will be the center of curvature for point 1. The distances $R'$ and $R''$ will tend to a common limit $R$ equal to the radius of curvature. Indeed, if points 1 and 1′ are close to each other, we can write that $\Delta\varphi \approx \Delta s/R'$ or $R' \approx \Delta s/\Delta\varphi$. In the limit when $\Delta\varphi \to 0$, this approximate equation will transform into the strict equation $R = \mathrm{d}s/\mathrm{d}\varphi$ coinciding with the definition of the radius of curvature [see Eq. (1.33)].

Let us now turn to the calculation of $\boldsymbol{a}_n$[see Eq. (1.30)]. According to Eq. $\varphi\dot{\boldsymbol{e}}_a = \dot{\varphi}\boldsymbol{e}_\perp$,

$$\dot{\boldsymbol{\tau}} = \frac{\mathrm{d}\varphi}{\mathrm{d}t}\boldsymbol{n} \tag{1.34}$$

where $\boldsymbol{n}$ is the unit vector of the normal to the trajectory with its sense in the direction of rotation of the vector $\boldsymbol{\tau}$ when a particle travels along the trajectory . The quantity $\mathrm{d}\varphi/\mathrm{d}t$ can be related to the radius of curvature of the trajectory and the speed of the particle $v$. It follows from Fig. 1.14 that

$$\Delta\varphi \approx \frac{\Delta s}{R'} = \frac{v'\Delta t}{R'}$$

where $\Delta\varphi$ = angle of rotation of the vector $\boldsymbol{\tau}$ during the time $\Delta t$ (coinciding with the angle between the perpendiculars $R'$ and $R''$); $v'$ average speed over the distance $\Delta s$.

Hence,

$$\frac{\Delta\varphi}{\Delta t} \approx \frac{v'}{R'}$$

In the limit when $\Delta t$ tends to zero, the approximate equation will become a strict one, the average speed $v'$ will transform into the instantaneous speed $v$ at point 1, and $R'$ will become the radius of curvature $R$. As a result, we get the equation

$$\frac{d\varphi}{dt} = \frac{v}{R} = vC \tag{1.35}$$

($C$ is the curvature). Hence, the rate of rotation of the velocity vector, as we assumed, is proportional to the curvature of the trajectory and the speed of a particle along its trajectory.

Using Eq. (1.35) in Eq. (1.34), we find that $\dot{\boldsymbol{\tau}} = (v/R)\boldsymbol{n}$. And at last, introducing this expression into Eq. (1.30), we arrive at the final formula for the normal acceleration:

$$\boldsymbol{a}_n = \frac{v^2}{R}\boldsymbol{n} \tag{1.36}$$

Thus, the acceleration vector when a particle travels along a plane curve is determined by the following expression:

$$\boldsymbol{a} = \boldsymbol{a}_\tau + \boldsymbol{a}_n = \dot{v}\boldsymbol{\tau} + \frac{v^2}{R}\boldsymbol{n} \tag{1.37}$$

The magnitude of the vector $\boldsymbol{a}$ is

$$a = \sqrt{a_\tau^2 + a_n^2} = \sqrt{\dot{v}^2 + \left(\frac{v^2}{R}\right)^2} \tag{1.38}$$

In rectilinear motion, the normal acceleration is absent. It must be noted that $\boldsymbol{a}_n$ vanishes at the inflection point of a curvilinear trajectory (at point *IP* in Fig. 1.15). At both sides of this point, the vectors $\boldsymbol{a}_n$ have different directions. The vector $\boldsymbol{a}_n$ cannot change in a jump. Its direction reverses smoothly, and it becomes equal to zero at the inflection point.

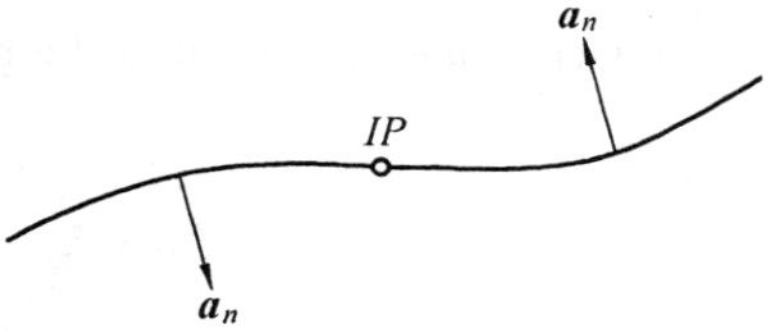

**Fig. 1.15**

Assume that a particle is traveling uniformly with an acceleration constant in magnitude. Since in uniform motion the magnitude of the velocity does not change, we have $\boldsymbol{a}_\tau = 0$, so that $\boldsymbol{a} = \boldsymbol{a}_n$. The constant magnitude of $\boldsymbol{a}_n$ signifies that $v^2/R$ const. Hence, we conclude that $R =$ const ($v =$ const because the motion is uniform). This means that the particle is traveling along a curve of constant curvature, i.e. a circle. Thus, when the acceleration of a particle is constant in magnitude and is directed at each moment of time at right angles to the velocity vector, the trajectory of the particle will be a circle.

## The Further Study:

### Circular Motion

The rotation of a body through a certain angle $\varphi$ can be given in the form of a straight line whose length is $\varphi$ and whose direction coincides with the axis about which the body is rotating. To indicate the direction of rotation about a given axis, it is related to the line depicting rotation by the right-hand screw rule: the line should be directed so that when looking along it (Fig.1.16) we see clockwise rotation (when rotating the head of a right-hand screw clockwise, we cause it to move away from us). We showed in Sec.1.2 (see Fig.1.4) that rotations through finite angles are not added by the parallelogram method and are therefore not vectors. Matters are different for rotations through very small angles $\Delta\varphi$. The distance traveled by any point of a body when rotated through a very small angle can be considered as a straight line (Fig. 1. 17). Consequently, two small circular motions $\Delta\varphi_1$ and $\Delta\varphi_2$ performed sequentially, as can be seen from the figure, result in the same displacement $\Delta\boldsymbol{r}_3 = \Delta\boldsymbol{r}_1 + \Delta\boldsymbol{r}_2$ of any point of the body as the circular motion $\Delta\varphi_3$ obtained from $\Delta\varphi_1$ and $\Delta\varphi_2$ by addition using the parallelogram method. Hence it follows that very small circular motions can be considered as vectors (we shall denote these vectors by $\Delta\varphi$ or $d\varphi$). The direction of the rotation vector is associated with the direction of rotation of a body. Consequently, $\mathrm{d}\varphi$ is not a true vector, but a pseudovector.

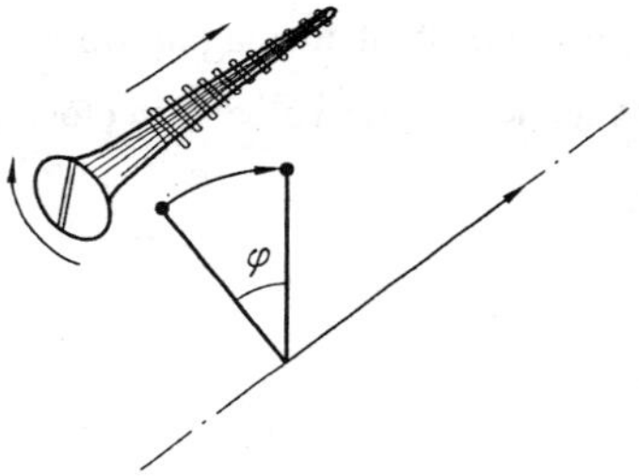


**Fig.1.16**

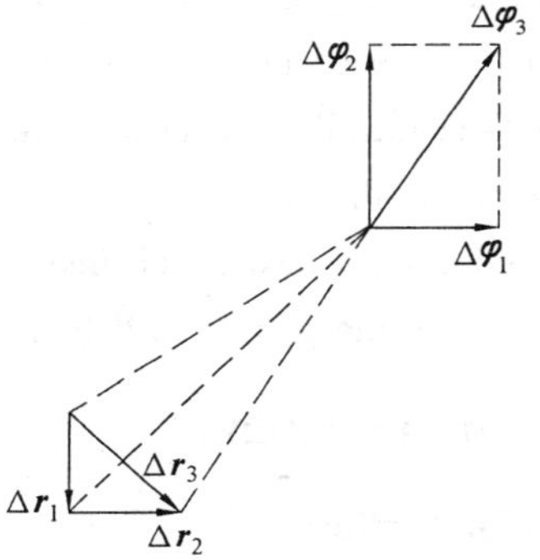


**Fig.1.17**

The vector quantity

$$\boldsymbol{\omega} = \lim_{\Delta t \to 0} \frac{\Delta \boldsymbol{\varphi}}{\Delta t} = \frac{\mathrm{d}\boldsymbol{\varphi}}{\mathrm{d}t} \tag{1.39}$$

(where $\Delta t$ is the time during which the circular motion $\Delta\boldsymbol{\varphi}$ is performed) is called the angular velocity of a body. The angular velocity $\boldsymbol{\omega}$ is directed along the axis about which, the body is totaling in a direction determined by the right-hand screw rule (Fig. 1.18) and is a pseudo vector. The magnitude of the angular velocity, i.e. the angular speed, equals $\mathrm{d}\varphi/\mathrm{d}t$. Circular motion at a constant angular velocity is called uniform.

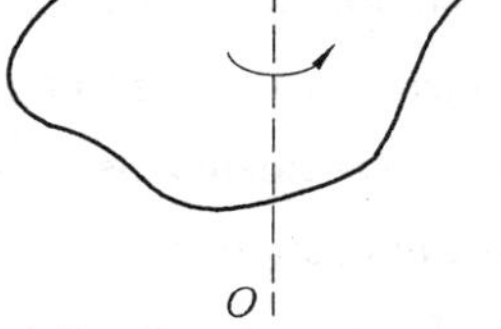

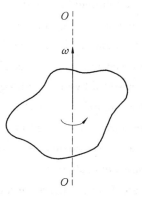

**Fig.1.18**

For uniform circular motion, we have $\omega = \varphi/t$, where $\varphi$ is the finite angle of rotation during the time $t$ (compare with $v = s/t$). Thus, in uniform circular motion, $\omega$ shows the angle through which a body rotates in unit time.

Uniform circular motion can be characterized by the period of revolution $T$. It is defined as the time during which a body completes one revolution, i.e. rotates through the angle $2\pi$ radians, or 360 degrees. Since the time interval $\Delta t = T$ corresponds to the angle of rotation $\Delta\varphi = 2\pi$, we have

$$\omega = \frac{2\pi}{T} \tag{1.40}$$

whence

$$T = \frac{2\pi}{\omega} \tag{1.41}$$

The number of revolutions in unit time $\nu$ is evidently equal to

$$\nu = \frac{1}{T} = \frac{\omega}{2\pi} \tag{1.42}$$

It follows from Eq. (1.42) that the angular velocity equals $2\pi$ multiplied by the number of revolutions per unit time:

$$\omega = 2\pi\nu \tag{1.43}$$

The concepts of the period of revolution and the number of revolutions per unit time can also be retained for non-uniform circular motion. Here, we must understand the instantaneous value of $T$ to signify the time during which a body would perform one revolution if it rotated uniformly with the given instantaneous value of the angular velocity, and $\nu$ to signify the number of revolutions which a body would complete in unit time in similar conditions.

The vector $\boldsymbol{\omega}$ may vary either as a result of a change in the speed of rotation of a body about its axis (in this case it changes in magnitude), or as a result of turning of the axis of rotation in space (in this case $\boldsymbol{\omega}$ changes in direction). Assume that during the time $\Delta t$ the vector $\boldsymbol{\omega}$ receives the increment $\Delta\boldsymbol{\omega}$.

The change in the angular velocity vector with time is characterized by the quantity

$$\boldsymbol{\alpha} = \lim_{\Delta t \to 0} \frac{\Delta \boldsymbol{\omega}}{\Delta t} = \frac{\mathrm{d}\boldsymbol{\omega}}{\mathrm{d}t} \tag{1.44}$$

called the angular acceleration. The latter, like the angular velocity, is a pseudovector.

Different points of a body in circular motion have different linear velocities $\boldsymbol{v}$. The velocity of each point continuously changes its direction. The speed $v$ is determined by the speed of rotation of the body $\boldsymbol{\omega}$ and the distance $R$ to the point being considered from the axis of rotation. Assume that during a small interval of time the body turned through the angle $\Delta\varphi$ (Fig. 1.19).

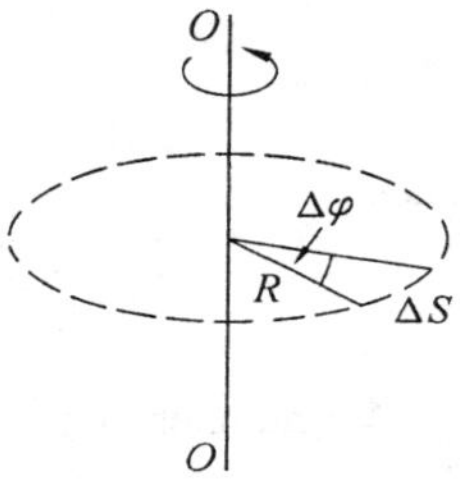


**Fig.1.19**

The point at the distance $R$ from the axis travels the path $\Delta s = R\Delta\varphi$. The linear speed of the point is

$$v = \lim_{\Delta t \to 0} \frac{\Delta s}{\Delta t} = \lim_{\Delta t \to 0} R\,\frac{\Delta \varphi}{\Delta t} = R \lim_{\Delta t \to 0} \frac{\Delta \varphi}{\Delta t} = R\,\frac{\mathrm{d}\varphi}{\mathrm{d}t} = Rw$$

Thus

$$v = \omega R \tag{1.45}$$

Equation (1.45) relates the linear and the angular speeds. Let us find an expression relating the relevant velocities $\boldsymbol{v}$ and $\boldsymbol{\omega}$. We shall determine the position of the point of the body being considered by the position vector $\boldsymbol{r}$ drawn from the origin of coordinates on the axis of rotation (Fig. 1.20). Examination of the figure shows that the vector

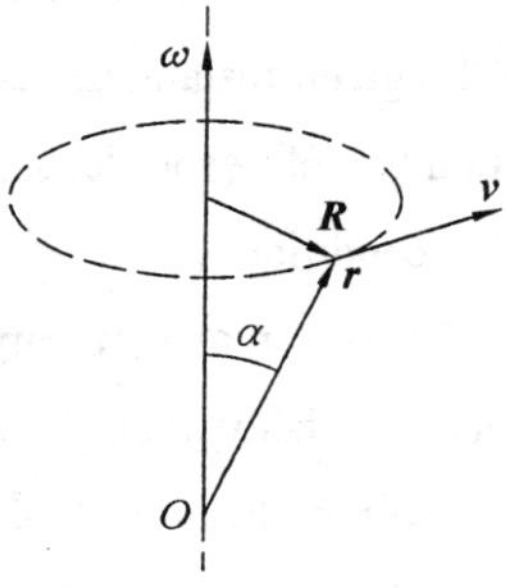


**Fig.1.20**

product $\boldsymbol{\omega r}$ coincides in direction with the vector $\boldsymbol{v}$ and its magnitude is $\omega r \sin\alpha = \omega R$. Hence,

$$\boldsymbol{v} = [\boldsymbol{\omega r}] \tag{1.46}$$

The normal acceleration of the points of a rotating body is $a_n = v^2/R$. Introducing into this expression the value of $\boldsymbol{v}$ from Eq. (1.45), we get

$$a_n = \omega^2 R \tag{1.47}$$

If we introduce the vector $\boldsymbol{R}$ drawn to the given point of the body from the axis of rotation at right angles to the latter (see Fig. 1.20), then Eq. (1.47) can be given a vector form:

$$\boldsymbol{a}_n = -\omega^2 \boldsymbol{R} \tag{1.48}$$

There is a minus sign in this formula because the vectors $\boldsymbol{a}_n$ and $\boldsymbol{R}$ have opposite directions.

Let us assume that the axis of rotation of a body does not turn 1 in space. According to Eq. (1.31), the magnitude of the tangential acceleration is $|\mathrm{d}v/\mathrm{d}t|$. Using equation (1.45) and taking into account that the distance to the point being considered from the axis of rotation $R =$ const, we can write

$$a_\tau = \left|\lim_{\Delta t\to 0}\frac{\Delta v}{\Delta t}\right| = \left|\lim_{\Delta t\to 0}\frac{\Delta(\omega R)}{\Delta t}\right| = \left|\lim_{\Delta t\to 0} R\frac{\Delta\omega}{\Delta t}\right| = R\left|\lim_{\Delta t\to 0}\frac{\Delta\omega}{\Delta t}\right| = R\alpha$$

where $\alpha$ is the magnitude of the angular acceleration. Consequently, the magnitude of the tangential acceleration is related to the magnitude of the angular acceleration as follows

$$a_\tau = \alpha R \tag{1.49}$$

Thus, the normal and tangential accelerations grow linearly with an increasing distance to a point from the axis of rotation.

## Summary of Key Terms

**Vector quantity** A quantity that has both magnitude and direction. Examples: force, velocity, acceleration, and momentum.

**Scalar quantity** A quantity that has magnitude, but no direction. examples: mass, volume, and speed.

**Vector** An arrow drawn to scale used to represent a vector quantity.

**Motion** A change of position.

**Speed** The distance traveled per time.

**Velocity** The rate of change position and the direct of the motion. Velocity is a vector quantity.

**Acceleration** Time rate of velocity: Acceleration = change in velocity/time it takes for change.

2

# Laws of Conservation

## 2.1 QUANTITIES OBEYING THE LAWS OF CONSERVATION

Bodies forming a mechanical system may interact with one another and with bodies not belonging to the given system. Accordingly the forces acting on the bodies of a system can be divided into internal and external ones. We shall define internal forces as the forces with which given body is acted upon by the other bodies of the system and external forces as those produced by the action of bodies not belonging to the system. If external forces are absent, the relevant system is called closed.

There are functions of the coordinates and velocities of the particles forming a system for closed systems that retain constant values upon motion. These functions are called motion integrals.

The number of motion integrals that can be formed for a system of N particles between which there are no rigid constraints is $6N - 1$. Only those of them are of interest to us that have the property of additively. This property consists in that the value of a motion integral for a system comprising parts whose interaction may be disregarded equals the sum of the values for each part. There are three additive motion integrals. The first is called **energy**, the second—**momentum**, and the third—**angular momentum**.

Thus, three physical quantities do not change in closed systems, namely, energy, momentum, and angular momentum. Accordingly there are three laws of conservation—that of energy conservation, that of momentum conservation, and that of angular momentum conservation. These laws are closely associated with the fundamental properties of space and time.

The conservation of energy is based on the uniformity of time i.e. the equivalence of all moments of time. The equivalence should be understood in the sense that the substitution of the moment of time $t_2$ for the moment $t_1$ without a change in the values of the coordinates and velocities of the particles does not change the mechanical properties of a system. This signifies that after such a substitution, the coordinates and velocities of the particles have the same values at any moment of time $t_2 + t$ as they would have had before the substitution at the moment $t_1 + t$.

The conservation of momentum is based on the **uniformity of space**, i.e. the identical properties of space at all points. This should be understood in the sense that a translation of a closed system from one place in space to another without changing the mutual arrangement and velocities of the particles does not change the mechanical properties of the system (it is assumed that the closed nature of the system is not violated at the new place).

Finally, the conservation of angular momentum is based on the isotropy of space, i.e. the identical properties of space in all directions. This should be understood in the sense that rotation of a closed system as a whole does not affect its mechanical properties.

The laws of conservation are a powerful means of research. It is often extremely difficult to accurately solve equations of motion. In these cases, the laws of conservation permit us to obtain numerous important data on how mechanical phenomena proceed without having to solve equations of motion. The laws of conservation do not depend on the nature of the acting forces. This is why they can help us obtain much important information on the behavior of mechanical systems even when the forces are unknown.

In the following sections, we shall obtain the laws of conservation on the basis of Newton's equations. It must be borne in mind, however, that the laws of conservation have a much more general nature than Newton's laws. The laws of conservation remain strictly correct even when Newton's laws (particularly the third one) are violated. We stress the fact that the laws of energy, momentum, and angular momentum conservation are accurate laws that are also strictly obeyed in the relativistic realm.

## 2.2 KINETIC ENERGY

Let us now pass over to finding the additive integrals of motion. We shall first consider the simplest system consisting of a single point particle.

The equation of motion of the particle is

$$m\dot{\boldsymbol{v}} = \boldsymbol{F} \tag{2.1}$$

Here $\boldsymbol{F}$ is the resultant of the forces acting on the particle. Multiplying Eq.(2.1) by the displacement of the particle $\mathrm{d}\boldsymbol{s} = \boldsymbol{v}\mathrm{d}t$, we get

$$m\boldsymbol{v}\dot{\boldsymbol{v}}\mathrm{d}t = \boldsymbol{F}\mathrm{d}\boldsymbol{s} \tag{2.2}$$

The product $\dot{\boldsymbol{v}}\mathrm{d}t$ is the increment of the velocity of the particle $\mathrm{d}\boldsymbol{v}$ during the time $\mathrm{d}t$. Accordingly

$$m\boldsymbol{v}\dot{\boldsymbol{v}}\mathrm{d}t = m\boldsymbol{v}\mathrm{d}\boldsymbol{v} = m\mathrm{d}\left(\frac{v^2}{2}\right) = \mathrm{d}\left(\frac{mv^2}{2}\right) \tag{2.3}$$

Performing such a substitution in Eq.(2.2), we arrive at the expression

$$d\left(\frac{mv^2}{2}\right) = \boldsymbol{F}d\boldsymbol{s} \tag{2.4}$$

If the system is closed, i.e. $\boldsymbol{F} = 0$, then $d(mv^2/2) = 0$, while the quantity

$$E_k = \frac{mv^2}{2} \tag{2.5}$$

itself remains constant. This quantity is called the kinetic energy of the particle. For an isolated particle, the kinetic energy is an integral of motion.

Multiplying the numerator and denominator of Eq.(2.5) by $m$ and taking into consideration that the product $mv$ equals the momentum $p$ of a body, the expression for the kinetic energy can be given the form

$$E_k = \frac{p^2}{2m} \tag{2.6}$$

If the force $\boldsymbol{F}$ acts on a particle, its kinetic energy does not remain constant. In this case in accordance with Eq.(2.4), the increment of the particle's kinetic energy during the time $dt$ equals the scalar product $\boldsymbol{F}d\boldsymbol{s}$ ($d\boldsymbol{s}$ is the displacement of the particle during the time $dt$). The quantity

$$dA = \boldsymbol{F}d\boldsymbol{s} \tag{2.7}$$

is called the work done by the force $\boldsymbol{F}$ over the path $ds$ ($ds$ is the magnitude of the displacement $d\boldsymbol{s}$). The scalar product (2.7) can be represented as the product of the projection of the force onto the direction of the displacement $F_s$ and the elementary distance $ds$. Consequently, we can write that

$$dA = F_s ds \tag{2.8}$$

It is clear from the above that work characterizes the change in energy due to the action of a force on a moving particle.

Let us integrate Eq.(2.4) along a certain trajectory from point 1 to point 2:

$$\int_1^2 \mathrm{d}\left(\frac{mv^2}{2}\right) = \int_1^2 \boldsymbol{F}\mathrm{d}\boldsymbol{s}$$

The left-hand side is the difference between the values of the kinetic energy at points 2 and 1, i.e. the increment of the kinetic energy along path 1 – 2. Taking this into account, we get:

$$E_{k,2} - E_{k,1} = \frac{mv_2^2}{2} - \frac{mv_1^2}{2} = \int_1^2 \boldsymbol{F}\mathrm{d}\boldsymbol{s} \tag{2.9}$$

The quantity

$$A = \int_1^2 \boldsymbol{F}\mathrm{d}\boldsymbol{s} = \int_1^2 F_s \mathrm{d}s \tag{2.10}$$

is the work of the force $\boldsymbol{F}$ over path 1 – 2. We shall sometimes denote this work by the symbol $A_{12}$ instead of $A$.

Thus, the work of the resultant of all the forces acting on a particle produces an increment of the particle's kinetic energy:

$$A_{12} = E_{k,2} - E_{k,1} \tag{2.11}$$

It follows from Eq. (2.11) that energy has the same dimension as work. Accordingly, energy is measured in the same units as work (see the following section).

## 2.3 WORK

Let us consider the quantity that we called work in greater detail. Equation (2.7) can be written in the form

$$\mathrm{d}A = \boldsymbol{F}\mathrm{d}\boldsymbol{s} = F\cos\alpha \cdot \mathrm{d}s \tag{2.12}$$

where $\alpha$ is the angle between the direction of the force and that of the displacement of the point of application of the force.

If the force and the direction of the displacement make an acute angle ($\cos\alpha > 0$), the work is positive. If the angle $\alpha$ is obtuse ($\cos\alpha < 0$), the work is negative. When $\alpha = \pi/2$, the work equals zero. This especially clearly shows that the concept of work in mechanics appreciably differs from our ordinary notion of it. In the ordinary meaning, any effort, particularly muscular strain, is always attended by work being done. For example, in order to hold a heavy load while standing still, and, moreover, to carry this load along a horizontal path, a porter spends much effort, i.e. "does work". The work as a mechanical quantity in these cases, however, equals zero.

Figure 2.1 is a plot of the projection of the force onto the direction of displacement $F_s$, as a function of the position of the particle on its trajectory (the axis of abscissas has been taken as the $s$-axis, the length of the part of this axis between points 1 and 2 equals the total length of the path). Examination of the figure shows that the elementary work $dA = F_s \times ds$ equals numerically the area of the hatched strip, while the work $A$ over path 1 – 2 equals numerically the area of the figure confined by the curve $F_s$, the vertical lines from points 1 and 2 and the $s$ – axis.

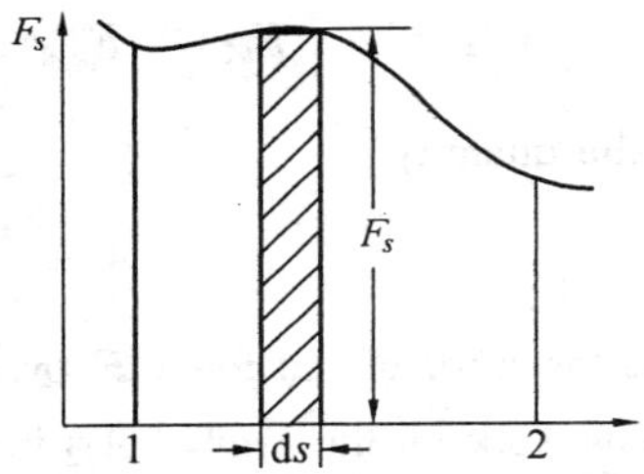


**Fig.2.1**

Let us use this result to find the work one in the deformation of a spring obeying Hooke's law. We shall begin with stretching of the spring. We shall do this very slowly so that the force $\boldsymbol{F}_{ext}$ which we act on the spring with may be considered equal in magnitude to the elastic force $\boldsymbol{F}_{el}$ all the time. Hence, $F_{x,ext} = -F_{x,el} = kx$, where $x$ is the elongation of the spring (Fig.2.2). A glance at the figure shows that the work required to cause the elongation $x$ of the spring is

$$A = \frac{kx^2}{2} \tag{2.13}$$

When the spring is compressed by the amount $x$, work of the same magnitude and sign is done as in stretching by $x$. The projection of the force $\boldsymbol{F}_{ext}$ in this case is negative ($\boldsymbol{F}_{ext}$ is directed to the left, $x$ grows to the right, see Fig.2.2), and all the $dx$'s are also negative. As a result, the product $F_{x,ext}\mathrm{d}x$ is positive.

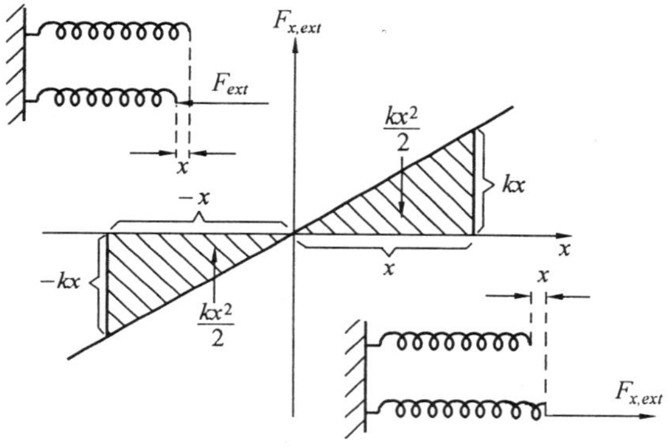


**Fig.2.2**

In a similar way, we can find an expression for the work done upon the elastic stretching or compression of a bar. According to Eq.(2.31), this work is

$$A = \frac{1}{2}\frac{ES}{l_0}(\Delta l)^2 = \frac{1}{2}ESl_0\left(\frac{\Delta l}{l_0}\right)^2 = \frac{1}{2}EV\varepsilon^2 \qquad (2.14)$$

where $V = Sl_0$ is the volume of the bar, and $\varepsilon = \Delta l / l_0$ is the relative e-longation.

Assume that several forces whose resultant is $\boldsymbol{F} = \sum_i \boldsymbol{F}_i$ act simultaneously on a body. It follows from the distributivity of a scalar product of vectors that the work $\mathrm{d}A$ done by the resultant force over the path $\mathrm{d}\boldsymbol{s}$ can be represented in the form

$$\mathrm{d}A = \left(\sum_i \boldsymbol{F}_i\right)\mathrm{d}\boldsymbol{s} = \sum_i \boldsymbol{F}_i\mathrm{d}\boldsymbol{s} = \sum_i \mathrm{d}A_i \qquad (2.15)$$

This signifies that the work of the resultant of several forces equals the algebraic sum of the work done by each force separately.

The elementary displacement $\mathrm{d}\boldsymbol{s}$ can be represented as $\boldsymbol{v}\mathrm{d}t$. We can therefore write the expression for the elementary work in the form

$$\mathrm{d}A = \boldsymbol{F}\boldsymbol{v}\mathrm{d}t \qquad (2.16)$$

The work done during the interval from $t_1$ to $t_2$ can thus be calculated by the formula

$$A = \int_{t_1}^{t_2} \boldsymbol{F}\boldsymbol{v}\mathrm{d}t \tag{2.17}$$

We have $\boldsymbol{F}\mathrm{d}\boldsymbol{s} = F\mathrm{d}s_F$, where $\mathrm{d}s_F$ is the projection of the elementary displacement $\mathrm{d}\boldsymbol{s}$ onto the direction of the force $\boldsymbol{F}$. The formula for work can therefore be written as follows:

$$\mathrm{d}A = F\mathrm{d}s_F \tag{2.18}$$

If the force has a constant magnitude and direction (Fig. 2.3), then the vector $\boldsymbol{F}$ in the expression for work may be put outside the integral. The result is

$$A = \boldsymbol{F}\int_1^2 \mathrm{d}\boldsymbol{s} = \boldsymbol{F}\boldsymbol{s} = Fs_F \tag{2.19}$$

where $\boldsymbol{s}$ is the vector of the displacement from point 1 to point 2, and $s_F$ is its projection onto the direction of the force.

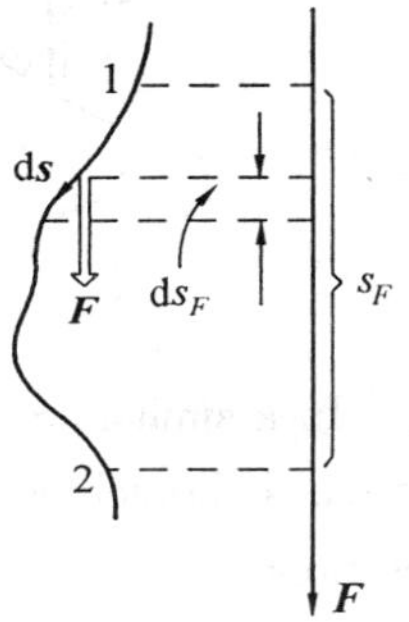

Fig.2.3

The work done in unit time is called power. If the work $\mathrm{d}A$ is done in the time $\mathrm{d}t$, then the power is

$$p = \frac{\mathrm{d}A}{\mathrm{d}t} \tag{2.20}$$

Taking $\mathrm{d}A$ as given by Eq. (2.16), we get the following expression for the power:

$$P = \boldsymbol{F}\boldsymbol{v} \tag{2.21}$$

according to which the power equals the scalar product of the force vector and the vector of the velocity with which the point of application of the force is moving.

## Units of Work and Power

The unit of work is the work done by a force equal to unit and acting in the direction of the displacement over unit distance. Consequently,

(1) in the SI system, the unit of work is the joule (J)—the work

done by a force of 1 N over a distance of 1 m;

(2) in the cgs system, the relevant unit is the erg—the work done by a force of 1 dyn over a distance of 1 cm;

(3) in the mkg(force)s system, the unit is the kilogramme(force) × m (kgf·m)—the work done by a force of 1 kgf over a distance of 1 m. The units of work are related as follows:

$$1\ \text{J} = 1\ \text{N} \times 1\ \text{m} = 10^5 \text{dyn} \times 10^2\ \text{cm} = 10^7\ \text{erg};$$

$$1\ \text{kgf}\cdot\text{m} = 1\ \text{kgf} \times 1\ \text{m} = 9.81\ \text{N} \times 1\ \text{m} = 9.81\ \text{J}.$$

The unit of power is the power at which per unit of work is done in unit time. The unit of power in the SI system is the watt (W) equal to one joule per second (J/s). The unit of power in the cgs system (erg/s) has no special name. The relation between the watt and the erg/s is $1\text{W} = 10^7 \text{erg/s}$.

The unit of power in the mkg(force)s system is the (metric) horsepower (hp), equal to 75 kgf·m/s, 1hp = 736W(do not confuse this unit with the British or U.S. horsepower equal to 550 ft – lb/s or 746W).

A system of prefixes is used, especially in the SI system, to denote multiples and submultiples of units. The names and symbols of these prefixes and the relevant factor by which the basic unit is multiplied are indicated in Table 2.1.

**Table 2.1 Prefixes for Multiples and Submultiples of Units**

| Name | Symbol | Factor by which unit is multiplied | Name | Symbol | Factor by which unit is multiplied |
|---|---|---|---|---|---|
| tera | T | $10^{12}$ | centi | c | $10^{-2}$ |
| giga | G | $10^{9}$ | milli | m | $10^{-3}$ |
| mega | M | $10^{6}$ | micro | μ | $10^{-6}$ |
| kilo | k | $10^{3}$ | nano | n | $10^{-9}$ |
| hecto | h | $10^{2}$ | pico | p | $10^{-12}$ |
| deca | da | $10^{1}$ | femto | f | $10^{-15}$ |
| deci | d | $10^{-1}$ | atto | a | $10^{-18}$ |

For example, the unit of work called the megajoule is equivalent to $10^6$ joules ($1\ MJ = 10^6\ J$), and the unit of power called the microwatt is equivalent to $10^{-6}$ watt ($1\ \mu W = 10^{-6}\ W$). Similarly, micrometer (formerly called the micron) is equivalent to $10^{-6}$ meter ($1\ \mu m = 10^{-6}\ m$), and $1\ pN = 10^{-12}\ N$.

## 2.4 CONSERVATIVE FORCES

If a particle is subjected to the action of other bodies at every point of space, the particle is said to be in a field of forces. For example, a particle near the Earth's surface is in the field of gravity forces — at every point of space the force $\boldsymbol{P} = m\boldsymbol{g}$ acts on it.

Let us consider as a second example the charged particle e in the electric field set up by the fixed-point charge $q$ (Fig. 2.4). A feature of this field is that the direction of the force acting on the particle at any point of space passes through a fixed center (the charge $q$), while the magnitude of the force depends only on the distance to this center, i.e. $F = F(r)$. A field of forces with such properties is called a central one.

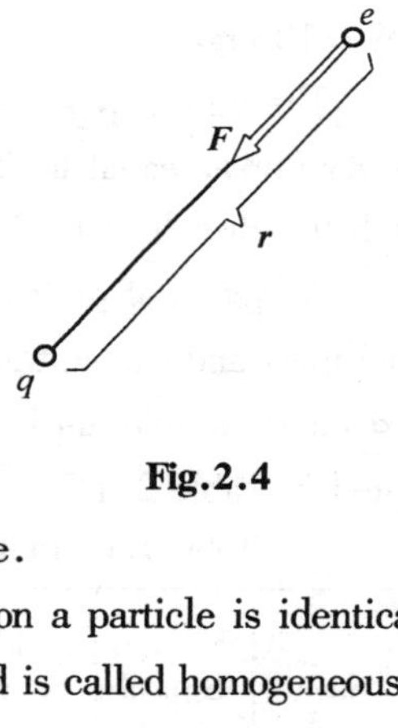

Fig.2.4

If at every point of a field the force acting on a particle is identical in magnitude and direction ($\boldsymbol{F}$ = const), the field is called homogeneous.

A field that changes with time is called non-stationary. A field that remains constant with time is called stationary.

For a stationary field, the work done on a particle by the forces of the field may depend only on the initial and final positions of the particle and not depend on the path along which the particle moved. Forces having such a property are called conservative.

It follows from the work of conservative forces being independent of the path that the work of such forces along a closed path equals zero. To prove this, let us divide an arbitrary closed path into two parts: path I along which a particle passes from point 1 to point 2, and path II along which the particle passes from point 2 to point 1 (Fig.2.5). We have chosen points 1 and 2 arbitrarily. The work along the entire closed path equals the sum of the work done on each of the parts:

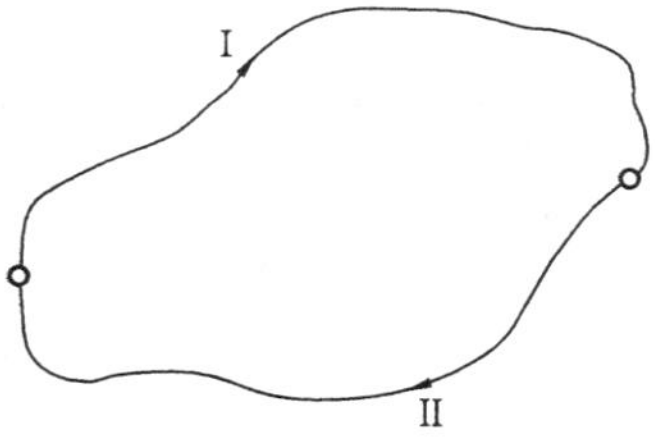


**Fig.2.5**

$$A = (A_{12})_{\mathrm{I}} + (A_{21})_{\mathrm{II}} \tag{2.22}$$

It is easy to see that the work $(A_{21})_{\mathrm{II}}$ differs from $(A_{21})_{\mathrm{II}}$ only in its sign. Indeed, reversing of the direction of motion results in $\mathrm{d}\boldsymbol{s}$ being replaced with $-\mathrm{d}\boldsymbol{s}$, and as a consequence the value of the integral $\int \boldsymbol{F}\mathrm{d}\boldsymbol{s}$ reverses its sign. Thus, Eq.(2.22) can be written in the form

$$A = (A_{12})_{\mathrm{I}} - (A_{21})_{\mathrm{II}}$$

and since the work does not depend on the path, i.e. $(A_{21})_{\mathrm{I}} = (A_{21})_{\mathrm{II}}$ we arrive at the conclusion that $A = 0$.

From the equality to zero of the work over a closed path, it is easy to obtain that the work $A_{21}$ is independent of the path. This can be done by reversing the above reasoning.

Thus, conservative forces can be defined in two ways: (1) as forces whose work does not depend on the path along which a particle passes from one point to another, and (2) as forces whose work along any closed path equal zero.

We shall prove that the force of gravity is conservative. This force at any point has the same magnitude and direction-vertically downward (Fig.2.6). Therefore, regardless of the path along which the particle moves (for example I or II in the figure), the work $A_{21}$ according to Eq.

(2.19) is determined by the expression

$$A_{12} = mgs_{12} = mg(s_{12})_{pr,g}$$

Inspection of Fig.2.6 shows that the projection of the vector $s_{12}$ onto the direction $g$ equals the difference between the heights $h_1 - h_2$. Hence, the expression for the work can be written in the form

$$A_{12} = mg(h_1 - h_2) \quad (2.23)$$

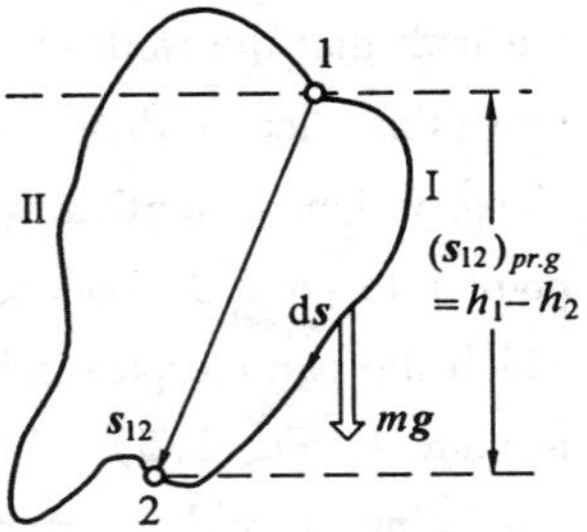


Fig.2.6

This expression obviously does not depend on the path. Hence it follows that the force gravity is conservative.

It is a simple matter to see that the same result is obtained for any stationary homogeneous field.

The force acting on a particle in a central field are also conservative. By Eq.(2.18) the elementary work over the path d$s$ (Fig.2.7) is

$$dA = F(r)ds_F$$

But the projection of d$s$ onto the direction of the force at a given point, i.e. onto the direction of the position vector $\boldsymbol{r}$, is d$r$—the increment of the distance from the particle to the force centre $O$, namely, $ds_F = dr$. Hence, $dA = F(r)dr$, and the work along the entire path is

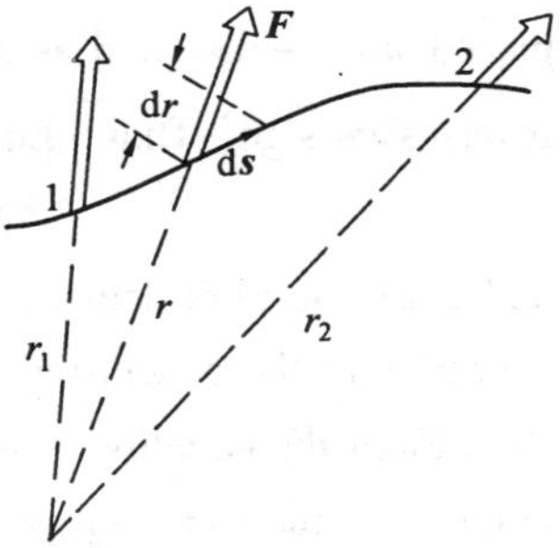


Fig.2.7

$$A_{12} = \int_{r_1}^{r_2} F(r)dr \qquad (2.24)$$

Equation (2.24) depends only on the form of the function $F(r)$ and on the value of $r_1$ and $r_2$. It does not depend in any way on the form of the trajectory whence it follows that the forces are conservative.

For our reader not to form the erroneous idea that any force depending only on the coordinates of a point is conservative, let us consider the following example. Assume that the components of a force are determined by the equations

$$F_x = ay, F_y = -ax, F_z = 0 \tag{2.25}$$

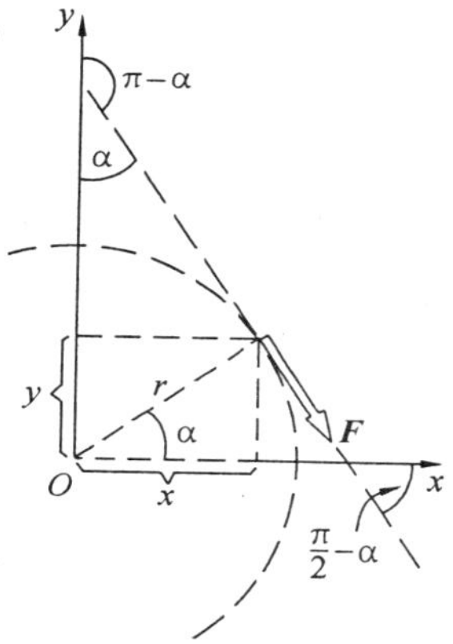

**Fig.2.8**

This force has a magnitude equal to $F = ar$, and is directed along a tangent to a circle of radius $r$ (Fig. 2.8). Indeed, as follows from the figure, for a force of such a magnitude and direction, we have

$$F_x = ar\cos(\frac{\pi}{2} - \alpha) = ar\sin\alpha = ar\frac{y}{r} = ay$$

$$F_y = ar\cos(\pi - \alpha) = -ar\cos\alpha = -ar\frac{x}{r} = -ax$$

which coincide with the values given by Eq. (2.25). Let us take a closed path in the form of a circle of radius $r$ with its center at the origin of coordinates. The work of the force along this path evidently equals $F \times 2\pi r = ar \times 2\pi r = 2\pi ar^2$, i.e. does not equal zero. Consequently, the force is not conservative.

Forces of friction are typical non-conservative ones. Since the force of friction $\boldsymbol{F}$ and the velocity of a particle $\boldsymbol{v}$ are directed oppositely, then the work of the force of friction on each part of the path is negative:

$$dA = \boldsymbol{F}d\boldsymbol{s} = \boldsymbol{F}\boldsymbol{v}dt = -fv dt = -Fds < 0$$

Therefore, the work along any closed path will also be negative (i.e. other than zero). Hence it follows that the forces of friction are not conservative.

It must be noted that a field of conservative forces is a particular case of a potential force field. A field of forces is called potential if it can be described with the aid of the function $V(x, y, z, t)$, whose gradient [see the following section, Eq. (2.31)] determines the force at each point of the field: $\boldsymbol{F} = \nabla V$[compare with Eq. (2.32)]. The function $V$ is called the potential function or the potential. When a potential does not depend explicitly on the time, i.e. $V = V(x, y, z)$, the potential field is stationary, and its forces are conservative. In this case

$$V(x, y, z) = - E_p(x, y, z)$$

where $E_p(x, y, z)$ is the potential energy of a particle.

For a non-stationary force field described by the potential $V(x, y, z, t)$, the potential and conservative forces cannot be considered identical.

## 2.5 POTENTIAL ENERGY IN AN EXTERNAL FORCE FIELD

Let us consider the case when the work of field forces does not depend on the path, but depends only on the initial and final position of a particle in the field. A value of a certain function $E_p(x, y, z)$ can be assigned to each point of the field such that the difference between the values of this function at points 1 and 2 will determine the work of the forces when the particle passes from the first point to the second one:

$$A_{12} = E_{p,1} - E_{p,2} \tag{2.26}$$

We can assign this function as follows. We take an arbitrary value of the function equal to $E_{p,0}$, for an initial point 0. We assign the value

$$E_p(P) = E_{p,0} + A_{p,0} \tag{2.27}$$

to any other point $P$. Here $A_{p,0}$ is the work done on a particle by the conservative forces when it is moved from point $P$ to point 0. Since the work is independent of the path, the value of $E_p(P)$ determined in this way will be unambiguous. It must be noted that the function $E_p(P)$ has the dimension of work (or energy).

In accordance with Eq. (2.27), the values of the function at points 1 and 2 are

$$E_{p,1} = E_{p,0} + A_{10}, E_{p,2} = E_{p,0} + A_{20}$$

Let us form the difference between these values and take into account that $A_{20} = -A_{02}$. As a result, we get

$$E_{p,1} - E_{p,2} = A_{10} - A_{20} = A_{10} + A_{02}$$

The sum $A_{10} + A_{02}$ gives the work done by the forces of the field when the particle moves from point 1 to point 2 along a trajectory passing through point 0. However, the work done to move the particle from point 1 to point 2 along any other trajectory (including one not passing through point 0) will be the same. Hence, the sum $A_{10} + A_{02}$ can be written simply in the form $A_{12}$. As a result, we get Eq. (2.26).

We can thus use the function $E_p$ to determine the work done on a particle by conservative forces along any path beginning at arbitrary point 1 and terminating at arbitrary point 2.

Assume that only conservative forces act on the particle. Consequently, the work done on the particle along path 1 – 2 can be represented in the form of Eq. (2.26). According to Eq. (2.11), this work produces an increment of the kinetic energy of the particle. We thus arrive at the equation

$$E_{k,2} - E_{k,1} = E_{p,1} - E_{p,2}$$

whence it follows that

$$E_{k,2} + E_{p,2} = E_{k,1} + E_{p,1}$$

The result obtained signifies that the quantity

$$E = E_k + E_p \qquad (2.28)$$

for a particle in the field of conservative forces remains constant, i.e. is an integral of motion.

It follows from Eq. (2.28) that $E_p$ is an addend in the motion integral having the dimension of energy. In this connection, the function $E_p$

$(x, y, z)$ is called the potential energy of a particle in an external force field. The quantity E equal to the sum of the kinetic and potential energies is called the total mechanical energy of the particle.

According to Eq. (2.26), the work done on a particle by conservative forces equals the decrement of the potential energy of the particle. We can say in a different way that work is done at the expense of the store of potential energy.

We can see from Eq. (2.27) that the potential energy is determined with an accuracy to a certain unknown additive constant $E_{p,0}$. This circumstance is of no significance, however, because all physical relations contain either the difference between the values of $E_p$ for two positions of a body, or the derivative of the function $E_p$, with respect to the coordinates. In practice, the potential energy of a body at a certain position is considered to equal zero, and the energy at other positions is taken with respect to this energy.

Knowing the form of the function $E_p(x, y, z)$, we can find the force acting on a particle at every point of a field. Let us consider the displacement of a particle parallel to the $x$-axis by the amount $dx$. This displacement is attended by work being done on the particle that is $dA = \boldsymbol{F} d\boldsymbol{s} = F_x dx$ (the displacement components $dy$ and $dz$ equal zero). According to Eq. (2.26), the same work can be represented as the decrement of the potential energy: $dA = -dE_p$. Equating the two expressions for the work, we obtain

$$F_x dx = - E_p$$

Whence

$$F_x = -\frac{dE_p}{dx}(y = \text{const}, z = \text{const})$$

The expression in the right-hand side is the derivative of the function $E_p(x, y, z)$ calculated on the assumption that the variables $y$ and $z$ remain constant, and only the variable $x$ changes. Such derivatives are called partial ones and are denoted, unlike derivative functions of one variable, by the symbol $\frac{\partial E_p}{\partial x}$. Consequently, the component of the force along the $x$-axis equal the partial derivative of the potential energy with respect to the variable $x$ taken with the opposite sign:

$$F_x = -\frac{\partial E_p}{\partial x}$$

Similar expressions are obtained for the components of the force along the $y$-axes and $z$-axes. Thus,

$$F_x = -\frac{\partial E_p}{\partial x}, F_y = -\frac{\partial E_p}{\partial y}, F_z = -\frac{\partial E_p}{\partial z} \tag{2.29}$$

Knowing its components, we can find the force vector:

$$\boldsymbol{F} = F_x\boldsymbol{e}_x + F_y\boldsymbol{e}_y + F_z\boldsymbol{e}_z = -\frac{\partial E_p}{\partial x}\boldsymbol{e}_x - \frac{\partial E_p}{\partial y}\boldsymbol{e}_y - \frac{\partial E_z}{\partial z}\boldsymbol{e}_z \tag{2.30}$$

A vector having the components

$$\frac{\partial E_p}{\partial x}, \quad \frac{\partial E_p}{\partial y}, \quad \frac{\partial E_p}{\partial z}$$

where $\varphi$ is a scalar function of the coordinates $x$, $y$, $z$, is called the gradient of the function $\varphi$ and is designated by the symbol grade $\varphi$ or $\nabla\varphi$ ($\nabla$ stands for the nabla operator). It follows from the definition of the gradient that

$$\nabla\varphi = \frac{\partial\varphi}{\partial x}\boldsymbol{e}_x + \frac{\partial\varphi}{\partial x}\boldsymbol{e}_y + \frac{\partial\varphi}{\partial x}\boldsymbol{e}_z \tag{2.31}$$

A comparison of Eqs. (2.30 and 2.31) show that the conservative force equals the gradient of the potential energy taken with the opposite sign:

$$\boldsymbol{F} = -\nabla E_p \tag{2.32}$$

Assume that a particle which the force (2.32) act on moves over the distance d$s$ having the components d$x$, d$y$, d$z$. The force does the work

$$\mathrm{d}A = \boldsymbol{F}\mathrm{d}\boldsymbol{s} = -\nabla E_p\mathrm{d}\boldsymbol{s} = \left(\frac{\partial E_p}{\partial x}\mathrm{d}x + \frac{\partial E_p}{\partial y}\mathrm{d}y + \frac{\partial E_p}{\partial z}\mathrm{d}z\right)$$

Taking into account that $dA = -dE_p$ we get the following expression for the increment of the function $E_p$:

$$dE_p = \frac{\partial E_p}{\partial x}dx + \frac{\partial E_p}{\partial y}dy + \frac{\partial E_p}{\partial z}dz \qquad (2.33)$$

An expression such as Eq. (2.33) is called the total differential of the relevant function.

The concept of the total differential plays a great part in physics. For this reason, we shall devote a few lines to it. The total differential of the single-valued function $f(x, y, z)$ s defined as the increment which this function receives in transition from a point with the coordinates $x$, $y$, $z$ to a neighboring point with the coordinates $x + dx$, $y + dy$, $z + dz$. By definition, this increment equals

$$df(x, y, z) = f(x + dx, y + dy, z + dz) - f(x, y, z)$$

and, consequently, is determined only by the values of the function at the initial and final points. Hence, it can not depend on path along which transition occurs. Let us take the broken line consisting of a segments $dx$, $dy$, $dz$ as such a parts (Fig. 2.9). On the segments $dx$ the function $f(x, y, z)$ behaves like a function of one variable $x$, and receives the increment $\frac{\partial f}{\partial x}dx$. Similarly, on the segments $dy$ and $dz$, the function receives the increments $\frac{\partial f}{\partial y}dy$ and $\frac{\partial f}{\partial z}dz$. The total increment of the function when passing from the initial point to the final one thus equals

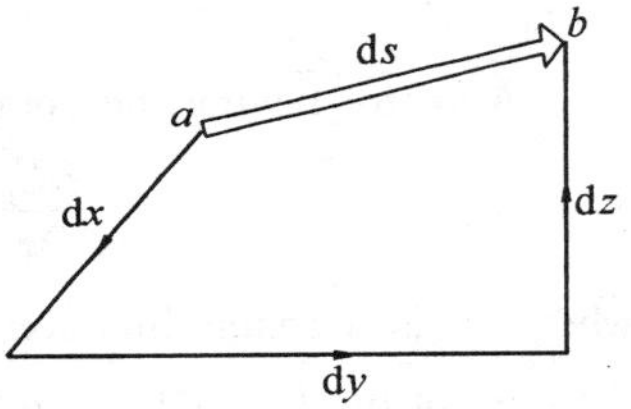

**Fig.2.9**

$$df(x, y, z) = \frac{\partial f}{\partial x}dx + \frac{\partial f}{\partial y}dy + \frac{\partial f}{\partial z}dz \qquad (2.34)$$

We have arrived at the expression for the total differential [compare with Eq. (2.33)].

Not any expression of the kind

$$P(x,y,z)\mathrm{d}x + Q(x,y,z)\mathrm{d}y + R(x,y,z)\mathrm{d}z$$

is a total differential of a certain function $f(x,y,z)$. Particularly the expression for the work done by the force whose projections are given by Eq. (2.25)

$$\mathrm{d}A = ay\mathrm{d}x - ax\mathrm{d}y \tag{2.35}$$

is not a total differential because there is no such function $E_p$ for which

$$-\frac{\partial E_p}{\partial x} = ay$$

and

$$-\frac{\partial E_p}{\partial x} = ax$$

[see Eq. (2. 25)]. Correspondingly, there is no function $E_p$ whose decrement would determine the work (2.35).

It follows from the above that only forces complying with the condition (2.32) can be conservative, i.e. such forces whose components along the coordinate axes equal the derivatives of a certain function $E_p(x, y, z)$ with respect to the relevant coordinates taken will the opposite sign. This function is the potential energy of a particle.

The concrete form of the function $E_p(x,y,z)$ depends on the nature of the force field. Let us find as an example the potential energy of a particle in a field of forces of gravity. According to Eq. (2.23), the work done on a particle by the forces of this field is

$$A_{12} = mg(h_1 - h_2)$$

On the other hand, according to Eq. (2.26),

$$A_{12} = E_{p,1} - E_{p,2}$$

Comparing these two expressions for the work, we arrive at the conclusion that the potential energy of a particle in a field of gravity forces is determined by the expression

$$E_p = mgh \tag{2.36}$$

where h is measured from an arbitrary level.

The zero of potential energy may be chosen arbitrarily. Therefore $E_p$ may have negative values. If we take the potential energy of a particle on the Earth's surface as zero, for example then the potential energy of a particle lying on the bottom of a pit with a depth of $h'$ will be $E_p = -mgh'$ (Fig. 2.10). It must be noted that the kinetic energy cannot be negative.

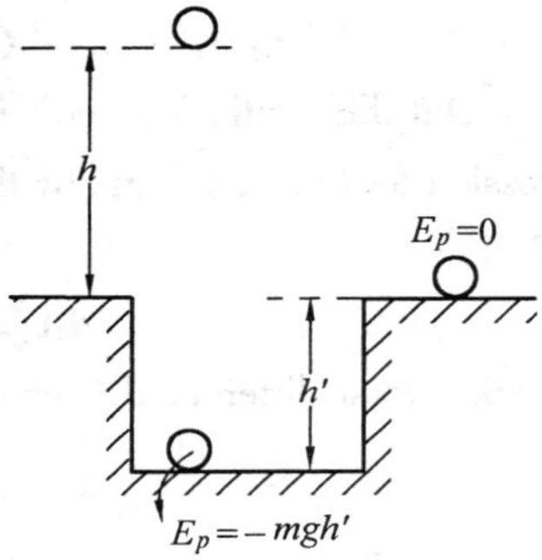

Fig.2.10

Assume that the non-conservative force $\boldsymbol{F}^*$ acts on a particle in addition to conservative forces. Hence, when the particle is moved from point 1 to point 2, the work done on it will be

$$A_{12} = \int_1^2 \boldsymbol{F}\mathrm{d}\boldsymbol{s} + \int_1^2 \boldsymbol{F}^*\mathrm{d}\boldsymbol{s} = A_{\text{cons}} + A_{12}^*$$

Where $A_{12}^*$ is the work of the non-conservative force. The work of the conservative forces $A_{\text{const}}$ can be represented as $E_{p,1} - E_{p,2}$. As a result, we find that

$$A_{12} = E_{p,1} - E_{p,2} + A_{12}^*$$

The total work of all the forces applied to the particle produces an increment of its kinetic energy [see Eq. (2.11)]. Consequently,

$$E_{k,2} - E_{k,1} = E_{p,1} - E_{p,2} + A_{12}^*$$

whence, taking into consideration that $E_k + E_p = E$, we get

$$E_2 - E_1 = A_{12}^* \tag{2.37}$$

The result obtained signifies that the work of non-conservative forces is spent on an increment of the total mechanical energy of a particle.

If the kinetic energy of a particle is the same in its final and initial positions (in particular, it equals zero), then the work of the nonconservative forces produces an increment of the potential energy of the particle:

$$A_{12}^{*} = E_{p,2} - E_{p,1} \tag{2.38}$$

$(E_{k,2} = E_{k,1})$. This relation is useful when finding the difference between the values of the potential energy.

Let us consider a system consisting of $N$ particles in the field of conservative forces when the particles do not interact with one another. Each of the particles has the kinetic energy

$$E_{k,i} = \frac{m_i v_i^2}{2} \ (i \text{ is the number of the particle})$$

and the potential energy

$$E_{p,i} = E_{p,i}(x_i, y_i, z_i)$$

Considering the $i$-th particle independently of the other particles, we can find that

$$E_i = E_{k,i} + E_{p,i} = \text{const}_i$$

Summating these equations for all the particles, we arrive at the relation

$$E = \sum_{i=1}^{N} E_i = \sum_{i-1}^{N} E_{k,i} + \sum_{i=1}^{N} E_{p,i} = \text{const} \tag{2.39}$$

This relation points to the additivity of the total mechanical energy for the system being considered.

According to Eq. (2.39), the total mechanical energy of a system of non-interacting particles on which only conservative forces act remains constant. This statement expresses the law of energy conservation for the above mechanical system.

If non-conservative forces $F_i^{*}$ act on particles in addition to conservative forces, the total energy of the system does not remain constant, and

$$E_2 - E_1 = \sum_{i=1}^{N} (A_{12}^{*})_i \tag{2.40}$$

where $(A_{12}^{*})_i$ is the work done by the non-conservative force applied to the $i$-th particle when it moves from its initial position to its final one.

We established at the end of the preceding section that the work of friction forces is always negative. Therefore, when such forces are present in a system, the total mechanical energy of the system diminishes (dissipates), transforming into non-mechanical forms of energy (for example, into the internal energy of bodies, or, as is customarily said, into heat). This process is called the dissipation of energy. Forces leading to the dissipation of energy are called dissipative. Thus, friction forces are dissipative. In general, forces that always act oppositely to the velocities of particles and therefore cause their retardation are called dissipative.

We shall note that non-conservative forces are not necessarily dissipative ones.

## The Further Study:

### Potential Energy of Interaction

Up to now, we treated systems of non-interacting particles. Now we shall pass over to the consideration of a system of two particles interacting with each other. Let $\boldsymbol{F}_{12}$ be the force with which the second particle acts on the first one, and $\boldsymbol{F}_{21}$ be the force with which the first particle acts on the second one. In accordance with Newton's third law, $\boldsymbol{F}_{12} = -\boldsymbol{F}_{21}$.

Let us introduce the vector $\boldsymbol{R}_{12} = \boldsymbol{r}_2 - \boldsymbol{r}_1$, where $\boldsymbol{r}_1$ and $\boldsymbol{r}_2$ are the position vectors of the particles (Fig. 2. 11). The distance between the particles equals the magnitude of this vector. Assume that the magnitudes of the forces $\boldsymbol{F}_{12}$ and $\boldsymbol{F}_{21}$ depend only on the distance $R_{12}$ between the particles, and that the forces are directed along the straight line connecting the particles. We know that this holds for forces of gravitational and Coulomb interactions.

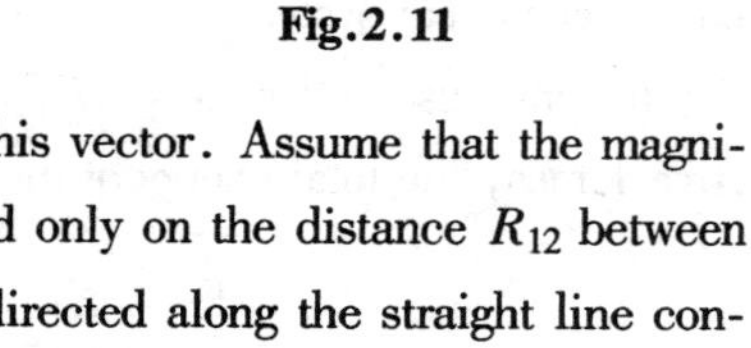

Fig.2.11

With these assumptions, the forces $\boldsymbol{F}_{12}$ and $\boldsymbol{F}_{21}$ can be represented in the form

$$F_{12} = -F_{21} = f(R_{12})e_{12} \tag{2.41}$$

where $e_{12}$ is the unit vector of $R_{12}$ (Fig.2.12), and $f(R_{12})$ is a certain function $R_{12}$ that is positive when the particles attract each other and negative when they repel each other.

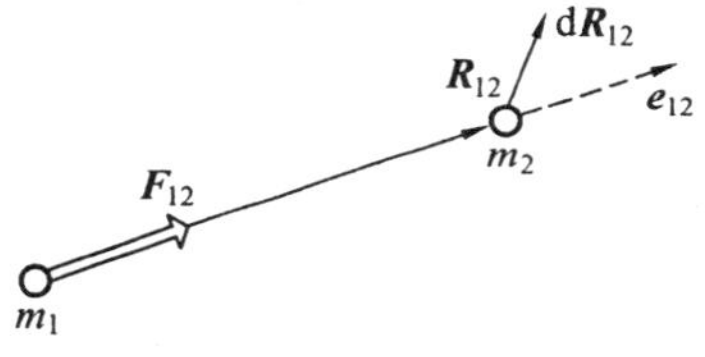


**Fig.2.12**

Considering our system to be closed (there are no external forces), let us write the equations of motion for our two particles:

$$m_1\dot{v}_1 = F_{12}, \quad m_2\dot{v}_2 = F_{21}$$

Let us multiply the first equation by $dr_1 = v_1dt$, the second by $dr_2 = v_2dt$, and add the resulting equations. We get

$$m_1v_1\dot{v}_1dt + m_2v_2\dot{v}_2dt = F_{12}dr_1 + F_{21}dr_2 \tag{2.42}$$

The left-hand side of this equation is the increment of the kinetic energy of the system during the time $dt$, and the right-hand side is the work of the internal forces during the same time. Taking into account that $F_{12} = -F_{21}$, we can write the right-hand side as follows:

$$dA_{int} = -F_{12}dr_1 + F_{21}dr_2 = -F_{12}d(r_2 - r_1) = -F_{12}dR_{12} \tag{2.43}$$

Introducing Eq.(2.41) for $F_{12}$ into the above equation, we get

$$dA_{int} = -f(R_{12})e_{12}dR_{12}$$

Examination of Fig.2.12 shows that the scalar product $e_{12}dR_{12}$ equals $dR_{12}$— the increment of the distance between the particles. Thus,

$$dA_{int} = -f(R_{12})dR_{12} \tag{2.44}$$

The expression $f(R_{12})dR_{12}$ can be considered as the increment of a certain function of $R_{12}$. Designating this function $E_p(R_{12})$, we arrive at the equation

$$f(R_{12})dR_{12} = dE_p(R_{12}) \tag{2.45}$$

Consequently,

$$dA_{int} = -dE_p \tag{2.46}$$

With a view to everything said above, Eq. (2.42) can be written in the form $dE_k = -dE_p$, or

$$dE = d(E_k + E_p) = 0 \tag{2.47}$$

whence it follows that the quantity $E = E_k + E_p$ for the closed system being considered remains unchanged. The function $E_p(R_{12})$ is the potential energy of interaction. It depends on the distance between the particles.

Let the particles move from their positions spaced $R_{12}^{(a)}$ apart to new positions spaced $R_{12}^{(b)}$ apart. In accordance with Eq. (2.46), the internal forces do the following work on the particles:

$$A_{ab,int} = -\int_a^b dE_p = E_p[R_{12}^{(a)}] - E_p[R_{12}^{(b)}] \tag{2.48}$$

It follows from Eq. (2.48) that the work of the forces (2.41) does not depend on the paths of the particles and is determined only by the initial and final distances between them (the initial and final configurations of the system). Forces of interaction of the form given by Eq. (2.41) are thus conservative.

If both particles move, the total energy of the system is

$$E = \frac{m_1 v_1^2}{2} + \frac{m_2 v_2^2}{2} + E_{p,ia}(R_{12}) \tag{2.49}$$

where $E_{p,ia}$ is the potential energy of interaction.

Assume that particle 1 is fixed at a certain point which we shall take as the origin of coordinates ($\boldsymbol{r}_1 = 0$). As a result, this particle will lose its ability to move, so that the kinetic energy will consist only of the single addend $\frac{m_2 v_2^2}{2}$. The potential energy will be a function only of $r_2$. Therefore, Eq. (2.49) becomes

$$E = \frac{m_2 v_2^2}{2} + E_{p,ia}(r_2) \tag{2.50}$$

If we consider the system consisting of only the single particle 2, then the function $E_{p,ia}$ will play the part of the potential energy of particle 2 in the field of the forces set up by particle 1. In essence, however, this function is the potential energy of interaction of particles 1 and 2. In general, the potential energy in an external field of forces is essentially the energy of interaction between the bodies of the system and those producing a force field that is external relative to the system.

Let us again turn to a system of two interacting free ("unfixed"), particles. If the external force $\boldsymbol{F}_1^*$ acts on the first particle in addition to the internal force, and the force $\boldsymbol{F}_2^*$ on the second particle, then the addends $\boldsymbol{F}_1^* \mathrm{d}\boldsymbol{r}_1$ and $\boldsymbol{F}_2^* \mathrm{d}\boldsymbol{r}_2$ will appear in the right-hand side of Eq. (2.42), and their sum will give the work of the external forces Equation (2.47) will correspondingly become

$$\mathrm{d}(E_k + E_{p,ia}) = \mathrm{d}A_{ext} \tag{2.51}$$

When the total kinetic energy of the particles remains constant (for example, equals zero), Eq. (2.51) becomes

$$\mathrm{d}E_{p,ia} = \mathrm{d}A_{ext} \tag{2.52}$$

(here $\mathrm{d}E_k = 0$). Integration of this equation from configuration a to configuration b yields

$$E_{p,ia}[R_{12}^{(b)}] - E_{p,ia}[R_{12}^{(a)}] = A_{ab,ext} \tag{2.53}$$

$(E_{k,a} = E_{k,b})$ [compare with Eq. (2.38)].

Let us extend the results obtained to a system of three interacting particles. In this case, the work of the internal forces is

$$\mathrm{d}A_{int} = (\boldsymbol{F}_{12} + \boldsymbol{F}_{13})\mathrm{d}\boldsymbol{r}_1 + (\boldsymbol{F}_{21} + \boldsymbol{F}_{23})\mathrm{d}\boldsymbol{r}_2 + (\boldsymbol{F}_{31} + \boldsymbol{F}_{32})\mathrm{d}\boldsymbol{r}_3 \tag{2.54}$$

Taking into account that $\boldsymbol{F}_{ik} = -\boldsymbol{F}_{ik}$, we can write Eq. (2.54) in the form

$$\begin{aligned}\mathrm{d}A_{int} &= \boldsymbol{F}_{12}\mathrm{d}(\boldsymbol{r}_2 - \boldsymbol{r}_1) - \boldsymbol{F}_{13}\mathrm{d}(\boldsymbol{r}_3 - \boldsymbol{r}_1) - \boldsymbol{F}_{23}\mathrm{d}(\boldsymbol{r}_3 - \boldsymbol{r}_2) = \\ &\quad - \boldsymbol{F}_{12}\mathrm{d}\boldsymbol{R}_{12} - \boldsymbol{F}_{13}\mathrm{d}\boldsymbol{R}_{13} - \boldsymbol{F}_{23}\mathrm{d}\boldsymbol{R}_{23}\end{aligned} \tag{2.55}$$

where

$$\boldsymbol{R}_{ik} = \boldsymbol{r}_k - \boldsymbol{r}_i$$

Let us assume that the internal forces can be represented in the form $\boldsymbol{F}_{ik} = f_{ik}(R_{ik}\boldsymbol{e}_{ik})$ [compare with Eq. (2.41)]. Hence,

$$\mathrm{d}A_{int} = -f_{12}(R_{12})\boldsymbol{e}_{12}\mathrm{d}\boldsymbol{R}_{12} - f_{13}(R_{13})\boldsymbol{e}_{13}\mathrm{d}\boldsymbol{R}_{13} - f_{23}(R_{23})\boldsymbol{e}_{23}\mathrm{d}\boldsymbol{R}_{23}$$

Each of the products $\boldsymbol{e}_{ik}\mathrm{d}\boldsymbol{R}_{ik}$ equals the increment of the distance between the corresponding particles $\mathrm{d}\boldsymbol{F}_{ik}$. Consequently,

$$\mathrm{d}A_{int} = -f_{12}(\boldsymbol{R}_{12})\mathrm{d}R_{12} - f_{13}(\boldsymbol{R}_{13})\mathrm{d}R_{13} - f_{23}(R_{23})\mathrm{d}\boldsymbol{R}_{23} =$$
$$-\mathrm{d}[E_{p,12}(R_{12}) + E_{p,13}(R_{13}) + E_{p,23}(R_{23})] = -\mathrm{d}E_{p,ia} \qquad (2.56)$$

Here,

$$E_{p,ia} = E_{p,12}(R_{12}) + E_{p,13}(R_{13}) + E_{p,23}(R_{23}) \qquad (2.57)$$

is the potential energy of interaction of the system . It consists of the energies of interaction of the particles taken in pairs.

Equating $\mathrm{d}E_k$ to the sum of the work $\mathrm{d}A_{int} = -\mathrm{d}E_{k,ia}$ and $\mathrm{d}A_{ext}$, we arrive at Eq. (2.51) in which by $E_{p,ia}$ we must understand Eq. (2.57).

The result obtained is easily generalized for a system with any number of particles. For a system of $N$ interacting particles, the potential energy of interaction consists of the energies of interaction of the particles taken in pairs:

$$E_{p,ia} = E_{p,12}(R_{12}) + E_{p,13}(R_{13}) + \cdots + E_{p,1N}(R_{1N}) +$$
$$E_{p,23}(R_{23}) + E_{p,24}(R_{24}) + \cdots + E_{p,2N}(R_{2N}) +$$
$$\cdots + E_{p,N-1,N}(R_{N-1,N}) \qquad (2.58)$$

This sum can be written as follows:

$$E_{p,ia} = \sum_{i<k} E_{p,ik}(R_{ik}) \qquad (2.59)$$

[note that in Eq. (2.58) the first subscript of each addend has a value smaller than the second one]. In connection with the fact that

$$E_{p,ik}(R_{ik}) = E_{p,ki}(R_{ki})$$

the energy of interaction can also be represented in the form

$$E_{p,ia} = \frac{1}{2}\sum_{i\neq k} E_{p,ik}(R_{ik}) \tag{2.60}$$

In the sums (2.59) and (2.60), the subscripts $i$ and $k$ take on values from 1 to $N$ with observance of the condition that $i < k$ or $i \neq k$.

Assume that a system consists of four particles, and that the first particle interacts only with the second one and the third particle only with the fourth one. The total energy of this system will be

$$\begin{aligned} E &= E_{k,1} + E_{k,2} + E_{k,3} + E_{k,4} + E_{p,12} + E_{p,34} = \\ &(E_{k,1} + E_{k,2} + E_{p,12}) + (E_{k,3} + E_{k,4} + E_{p,34}) = \\ &E' + E'' \end{aligned} \tag{2.61}$$

Here is $E'$ the total energy of the subsystem formed by particles 1 and 2, and $E''$ is the total energy of the subsystem formed by particles 3 and 4. In accordance with our assumption, there is no interaction between the subsystems. Equation(2.61) proves additivity of energy.

In conclusion, let us find the form of the function $E_{p,ia}$ for the case when the force of interaction is inversely proportional to the square of the distance between the particles:

$$f(R_{12}) = \frac{\alpha}{R_{12}^2} \tag{2.62}$$

($\alpha$ is a constant). We remind our reader that for attraction between the particles $\alpha > 0$, and for repulsion between them $\alpha < 0$ [see the text following Eq. (2.41)].

$$\mathrm{d}E_{p,ia} = f(R_{12})\mathrm{d}R_{12} = \frac{\alpha}{R_{12}^2}\mathrm{d}R_{12}$$

In accordance with Eq. (2.45)

$$\mathrm{d}E_{p,ia} = f(R_{12})\mathrm{d}R_{12} = \frac{\alpha}{R_{12}^2}\mathrm{d}R_{12}$$

Integration yields

$$E_{p,ia} = -\frac{\alpha}{R_{12}} + \text{const} \tag{2.63}$$

Like the potential energy in an external field of forces, the potential energy of interaction is determined with an accuracy up to an arbitrary additive constant. It is usually assumed that when $R_{12} = \infty$, the potential energy becomes equal to zero [at such a distance the force (2.62) becomes equal to zero – the interaction between the particles vanishes]. Hence, the additive constant in Eq.(2.63) vanishes, and the expression for the potential energy of interaction acquires the form

$$E_{p,ia} = -\frac{\alpha}{R_{12}} \tag{2.64}$$

In accordance with Eq.(2.53), the following work must be done to move the particles away from each other from the distance $R_{12}$ to infinity without changing their velocities:

$$A_{ext} = E_{p,ia,\infty} - E_{p,ia}(R_{12})$$

Introduction of the corresponding values of the function (2.64) leads to the expression

$$A_{ext} = 0 - \left(-\frac{\alpha}{R_{12}}\right) = \frac{\alpha}{R_{12}} \tag{2.65}$$

When the particles are attracted to each other, we have $\alpha > 0$; accordingly, positive work must be done to move the particles away from each other.

Upon repulsion of the particles from each other, $\alpha < 0$, and the work (2.65) is negative. This work has to be done to prevent the particles that are repelling each other from increasing their velocity.

## Summary of Key Terms

**Force** A quantity capable of producing motion or a change in motion.

**Net force** The equivalent or resultant force of two or more forces.

**Newton's first law of motion (law of inertia)** An object remains at rest or in uniform motion with a constant velocity unless acted upon by a net unbalanced force.

**Inertia** The property if matter that resists changes in motion. Mass is a measure of inertia.

**Newton's second law of motion** $F = ma$. Relates force to acceleration.

**Friction** The force that opposes the relative motion of contacting media.

**Mass** The quantity of matter in an object. More specifically, it is the measurement of the inertia or sluggishness that an object exhibits in response to any effort made to start it, or change in any way its state of motion.

**Weight** The force due to gravity on an object.

**Free fall** An object falling with only gravity acting on it.

**Terminal velocity** The maximum constant velocity that a non-freely falling object reaches when a retarding force balances the gravitation force and the object's acceleration is zero.

**Momentum** The product of the mass of an object and its velocity.

**Impulse** The product of the force acting on an object and the tine during which it acts.

**Elastic collision** A collision in which colliding objects rebound without lasting deformation or the generation of heat.

**Inelastic collision** A collision in which the colliding objects become distorted and generated heat during the collision.

**Power** The time rate of work $P = W/t$.

**Energy** The property of a system that enables it to do work.

# 3

# Mechanics of a Rigid Body

## 3.1 MOTION OF A BODY

In Sec. 1.1, we acquainted ourselves with the two fundamental kinds of motion of a rigid body translation and rotation.

In translation, all the points of a body receive displacements equal in magnitude and direction during the same time interval. Consequently, the velocities and accelerations of all the points are identical at every moment of time. It is therefore sufficient to determine the motion of one of the points of a body (for example, of its center of mass) to completely characterize the motion of the entire body.

In rotation, all the points of a rigid body move along circles whose centers are on a single straight line called the axis of rotation. To describe rotation, we must set the position of the axis of rotation in space and the angular velocity of the body at each moment of time.

Any motion of a rigid body can be represented as the superposition of the two fundamental kinds of motion indicated above. We shall show this for plane motion, i.e. motion when all the points of a body move in parallel planes. An example of plane motion is the rolling of a cylinder along a plane (Fig.3.1).

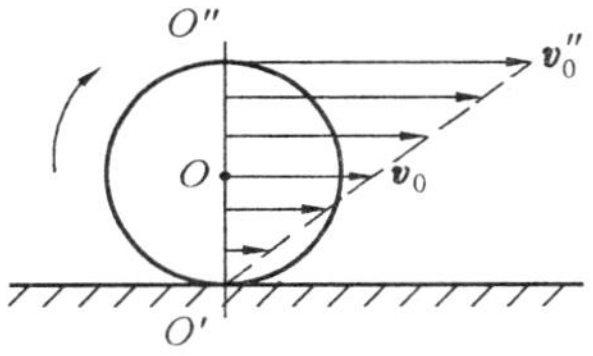


**Fig.3.1**

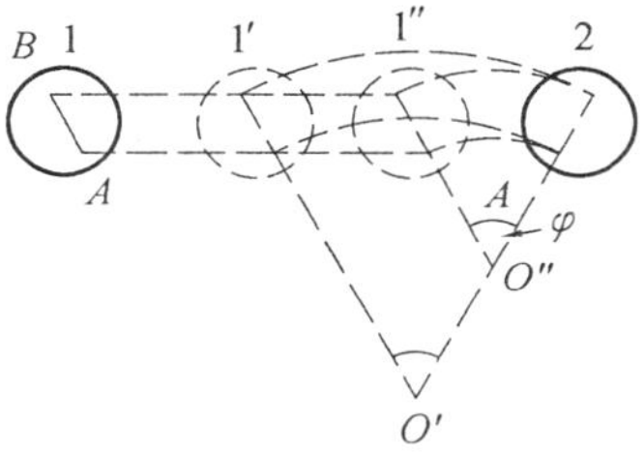


**Fig.3.2**

The arbitrary displacement of a rigid body from position 1 to position 2 (Fig.3.2) can be represented as the sum of two displacements – translation from position 1 to position 1′ or 1″ and rotation about the axis $O'$ or the axis $O''$. It is quite obvious that such a division of a displacement into translation and rotation can be performed in an infinite multitude of ways, but in any case rotation occurs through the same angle $\varphi$.

In accordance with the above, the elementary displacement of a point of a body $\mathrm{d}\boldsymbol{s}$ can be resolved into two displacements the translational one $\mathrm{d}\boldsymbol{s}_{\mathrm{tr}}$ and the "rotational" one $\mathrm{d}\boldsymbol{s}_{\mathrm{rot}}$:

$$\mathrm{d}\boldsymbol{s} = \mathrm{d}\boldsymbol{s}_{\mathrm{tr}} + \mathrm{d}\boldsymbol{s}_{\mathrm{rot}}$$

where $\mathrm{d}\boldsymbol{s}_{\mathrm{tr}}$ is the same for all the points of the body. This resolution of the displacement $\mathrm{d}\boldsymbol{s}$, as we have seen, can be performed in different ways. In each of them, the rotational displacement $\mathrm{d}\boldsymbol{s}_{\mathrm{rot}}$ is performed by rotation of the body through the same angle $\mathrm{d}\varphi$ but relative to different axes, whereas $\mathrm{d}\boldsymbol{s}_{\mathrm{tr}}$ and $\mathrm{d}\boldsymbol{s}_{\mathrm{rot}}$ are different.

Dividing $\mathrm{d}\boldsymbol{s}$ by the corresponding time interval $\mathrm{d}t$, we get the velocity of a point:

$$\boldsymbol{v} = \frac{\mathrm{d}\boldsymbol{s}}{\mathrm{d}t} = \frac{\mathrm{d}\boldsymbol{s}_{\mathrm{tr}}}{\mathrm{d}t} + \frac{\mathrm{d}\boldsymbol{s}_{\mathrm{rot}}}{\mathrm{d}t} = \boldsymbol{v}_0 + \boldsymbol{v}'$$

where $\boldsymbol{v}_0$ = velocity of translation, which is the same for all the points of a body; $\boldsymbol{v}'$ = velocity due to rotation, which is different for different points of the body.

Thus, the plane motion of a rigid body can be represented as the sum of two motions translation with the velocity $\boldsymbol{v}_0$ and rotation with the angular velocity $\boldsymbol{\omega}$ (the vector to in Fig.3.1 is directed at right angles to the plane of the drawing, beyond it). Such a representation of complex motion can be accomplished in many ways differing in the values of $\boldsymbol{v}_0$ and $\boldsymbol{v}'$, but corresponding to the same angular velocity $\boldsymbol{\omega}$. For example, the motion of a cylinder rolling without slipping along a plane (Fig.3.1) can be represented either as translation with the velocity $\boldsymbol{v}_0$ and simultaneous rotation with the angular velocity $\boldsymbol{\omega}$ about the axis $O$, or as translation with the velocity $\boldsymbol{v}''_0 = 2\boldsymbol{v}_0$ and rotation with the same angular velocity $\boldsymbol{\omega}$ about the axis $O''$, or, finally, as only rotation, again with the same angular velocity $\boldsymbol{\omega}$ about the axis $O'$.

Assuming that the reference frame relative to which we are considering the complex motion of a rigid body is stationary, the motion of the body can be represented as rotation with the angular velocity $\boldsymbol{\omega}$ in a reference frame moving translationally with the velocity $\boldsymbol{v}_0$ relative to the stationary frame.

The linear velocity $\boldsymbol{v}'$ of a point with the position vector $\boldsymbol{r}$ due to rotation of a rigid body is $\boldsymbol{v}' = [\boldsymbol{\omega r}]$. Consequently, the velocity of this point in complex motion can be represented in the form

$$\boldsymbol{v} = \boldsymbol{v}_0 + [\boldsymbol{\omega r}] \tag{3.1}$$

An elementary displacement of a rigid body in plane motion can always be represented as rotation about an axis called the instantaneous axis of rotation. This axis may be either inside the body or outside it. The position of the instantaneous axis of rotation relative to a fixed reference frame and relative to the body itself, generally speaking, changes with time. For a rolling cylinder (Fig.3.1), the instantaneous axis $O'$ coincides with the line of contact of the cylinder with the plane. When the cylinder rolls, the instantaneous axis moves both along the plane (i.e. relative to a fixed reference frame) and along the surface of the cylinder.

The velocities of all the points of the body for each moment of time can be considered as due to rotation about the corresponding instantaneous axis. Consequently, plane motion of a rigid body can be considered as a number of consecutive elementary rotations about instantaneous axes.

In non-planar motion, an elementary displacement of a body can be represented as rotation about an instantaneous axis only if the vectors $\boldsymbol{v}_0$ and $\boldsymbol{\omega}$ are mutually perpendicular. If the angle between these vectors differs from $\pi/2$, the motion of the body at each moment of time will be the superposition of two motions—rotation about a certain axis, and translation along this axis.

## 3.2 MOTION OF THE CENTER OF MASS OF A BODY

By dividing a body into elementary masses $m_i$, we can represent it as a system of point particles whose mutual arrangement remains unchanged. Any of these elementary masses may be acted upon both by internal forces due to its interaction with other elementary masses of the body being considered, and by external forces. For example, if a body is in the field of the Earth's gravitational forces, each elementary mass of the body $m_i$ will experience an external force equal to $m_i\boldsymbol{g}$.

Let us write the equation of Newton's second law for each elementary mass:

$$m_i\boldsymbol{a}_i = \boldsymbol{f}_i + \boldsymbol{F}_i \tag{3.2}$$

where $\boldsymbol{f}_i$ is the resultant of all the internal forces, and $\boldsymbol{F}_i$ the resultant of all the external forces applied to the given elementary mass. Summation of Eq. (3.2) for all the elementary masses yields

$$\sum m_i\boldsymbol{a}_i = \sum \boldsymbol{f}_i + \sum \boldsymbol{F}_i \tag{3.3}$$

The sum of all the internal forces acting in a system, however, equals zero. Hence, Eq. (3.3) can be simplified as follows:

$$\sum m_i \boldsymbol{a}_i = \sum \boldsymbol{F}_i \tag{3.4}$$

Here the resultant of all the external forces acting on the body is in the right-hand side.

The sum in the left-hand side of Eq. (3.4) can be replaced with the product of the mass of the body $m$ and the acceleration of its center of mass (center of inertia) $\boldsymbol{a}_c$. We have

$$\sum m_i \boldsymbol{r}_i = m\boldsymbol{r}_c$$

Differentiating this relation twice with respect to time and taking into account that $\ddot{\boldsymbol{r}}_i = \boldsymbol{a}_i$ and $\ddot{\boldsymbol{r}}_c = \boldsymbol{a}_c$, we can write

$$\sum m_i \boldsymbol{a}_i = m\boldsymbol{a}_c \tag{3.5}$$

Comparing Eq. (3.4) and (3.5), we arrive at the equation

$$m\boldsymbol{a}_c = \sum \boldsymbol{F}_{ext} \tag{3.6}$$

which signifies that the center of mass of a rigid body moves in the same way as a point particle of a mass equal to that of the body would move under the action of all the forces applied to the body.

Equation (3.6) permits us to find the motion of the center of mass of a rigid body if we know the mass of the body and the forces acting on it. For translation, this equation will determine the acceleration not only of the center of mass, but also of any other point of the body.

## 3.3 ROTATION OF A BODY ABOUT A FIXED AXIS

Let us consider a rigid body that can rotate about a fixed vertical axis (Fig. 3.3). We shall confine the axis in bearings to prevent its displacements in space. The flange *Fl* resting on the lower bearing prevents motion of the axis in a vertical direction.

A perfectly rigid body can be considered as a system of particles (point particles) with constant distances between them. Equation (a), i.e.

$$\frac{\mathrm{d}}{\mathrm{d}t}\boldsymbol{L} = \sum \boldsymbol{M}_{ext} \qquad (\mathrm{a})$$

holds for any system of particles, including a rigid body. In the latter case, $\boldsymbol{L}$ is the angular momentum of the body. The right-hand side of Eq. (a) is the sum of the moments of the external forces acting on the body.

Let us take point $O$ on the axis of rotation and characterize the position of the particles forming the body with the aid of position vectors $\boldsymbol{r}$ drawn from this point (Fig. 3.3 depicts the $i$-th particle of mass $m_i$). The angular momentum of the $i$ – th particle relative to point $O$ is

$$\boldsymbol{L}_i = [\boldsymbol{r}_i, m_i\boldsymbol{v}_i] = m_i[\boldsymbol{r}_i, \boldsymbol{v}_i] \qquad (3.7)$$

The vectors $\boldsymbol{r}_i$ and $\boldsymbol{v}_i$ are mutually perpendicular for all the particles of the body. Therefore, the magnitude of the vector $\boldsymbol{L}_i$[Eq. 3.7] is

$$L_i = m_i r_i v_i = m_i r_i \omega R_i \qquad (3.8)$$

The direction of the vector $\boldsymbol{L}_i$ is shown in Fig. 3.4. It must be noted that the "length" of the vector $\boldsymbol{L}_i$, according to Eq. (3.8) is proportional to the velocity of rotation of the body $\boldsymbol{\omega}$. The direction of the vector $\boldsymbol{L}_i$, however, is independent of $\boldsymbol{\omega}$. The vector $\boldsymbol{L}_i$ is in a plane passing through the axis of rotation and the particle $m$, and is perpendicular to $\boldsymbol{r}_i$.

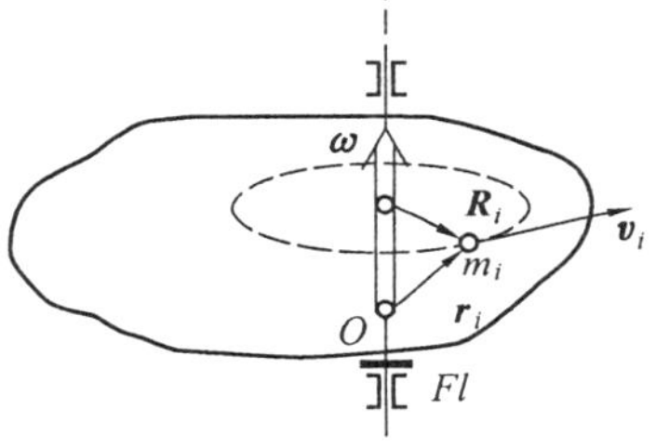


**Fig. 3.3**

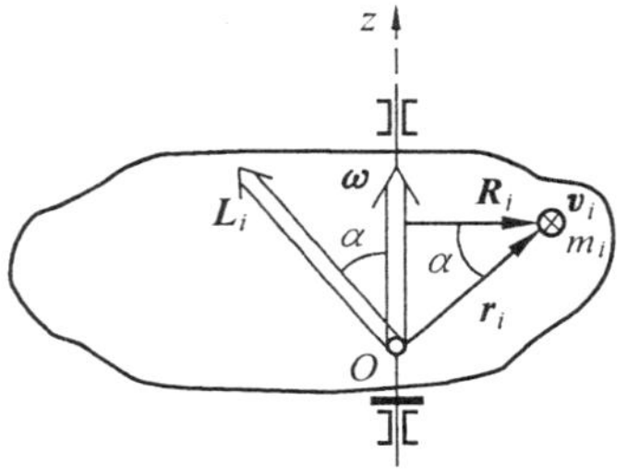


**Fig. 3.4**

The projection of the vector $\boldsymbol{L}_i$, onto the axis of rotation (the $z$ – axis), as can be seen from Fig. 3.4, is [see Eq. 3.8]

$$L_{zi} = L_i\cos\alpha = m_i r_i \omega R_i \cos\alpha = m_i(r_i\cos\alpha)R_i\omega = m_i R_i^2\omega \tag{3.9}$$

It is not difficult to see that for a homogeneous body which is symmetrical relative to the axis of rotation (for a homogeneous body of revolution), the directions of the total angular momentum (equal to $\sum L_i$ ) and of $\boldsymbol{\omega}$ along the axis of rotation are the same (Fig. 3.5). Indeed, in this case, the body can be divided into pairs of symmetrically arranged particles of equal mass (two pairs of particle are show in the figure—$m_i - m'_i$ and $m_k - m'_k$.) The sum of angular momenta of each pair (in the figure $\boldsymbol{L}'_i + \boldsymbol{L}''_i$ and) is directed along the vector $\boldsymbol{L}_k + \boldsymbol{L}'_k$ is directed along the vector $\boldsymbol{\omega}$. Hence, the total angular momentum $\boldsymbol{L}$ will also coincide in direction with $\boldsymbol{\omega}$. The magnitude of the vector $\boldsymbol{L}$ in this case equals the sum of the projections of the momenta $\boldsymbol{L}_i$ onto the $z$-axis. Taking Eq. (3.9) into account, we get the following expression for the magnitude of the angular momentum of a body:

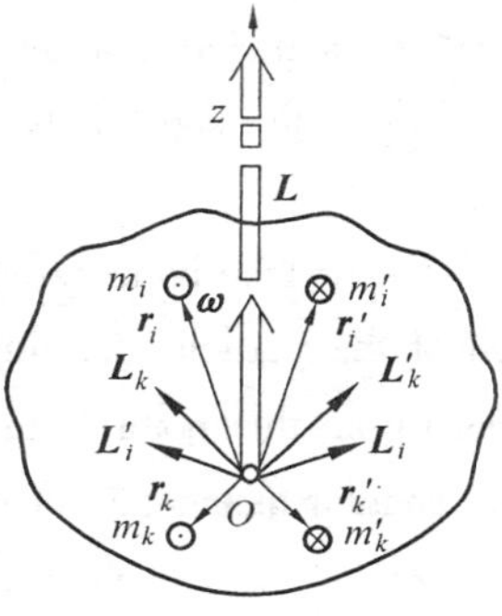

Fig.3.5

$$L = \sum L_{zi} = \omega \sum m_i R_i^2 = I\omega \tag{3.10}$$

The quantity $I$ equal to the sum of the products of the elementary masses and the squares of their distances from a certain axis is called the rotational inertia or the moment of inertia of the body relative to the given axis:

$$I = \sum m_i R_i^2 \tag{3.11}$$

Summation is performed over all the elementary masses $m_i$ into which the body was mentally divided.

With a view to the fact that the vectors $\boldsymbol{L}$ and $\boldsymbol{\omega}$ have identical directions, we can write Eq. (3.10) as follows:

$$L = I\omega \tag{3.12}$$

We remind our reader that we have obtained this relation for a homogeneous body rotating about an axis of symmetry. In the general case, as we shall see below, Eq.(3.12) is not obeyed.

For an asymmetrical (or non-homogeneous) body, the angular momentum $\boldsymbol{L}$, generally speaking, does not coincide in direction with the vector $\boldsymbol{\omega}$. The dash line in Fig.3.6 shows the part of an asymmetrical homogeneous body that is symmetrical relative to the axis of rotation. The total angular momentum of this part, as we have established above, is directed along $\boldsymbol{\omega}$. The momentum $\boldsymbol{L}_i$ of each particle not belonging to the symmetrical part deviates to the right from the axis of rotation (in a plane figure). Consequently, the total angular momentum of the entire body will also deviate to the right (Fig.3.7). Upon rotation of the body, the vector $\boldsymbol{L}$ rotates together with it, describing a cone. During the time $\mathrm{d}t$, the vector $\boldsymbol{L}$ receives the increment $\mathrm{d}\boldsymbol{L}$, which according to Eq.(a) equals

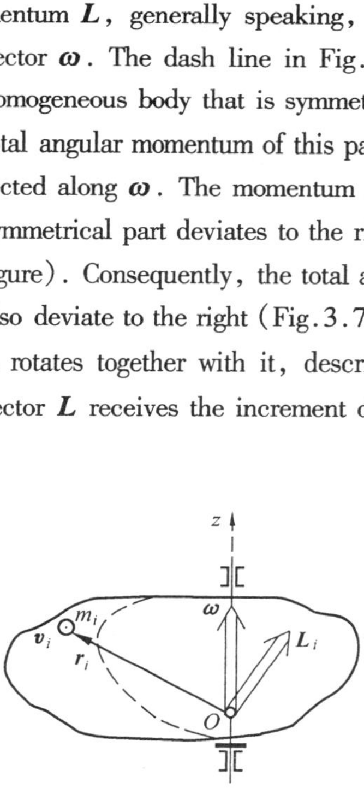


**Fig.3.6**

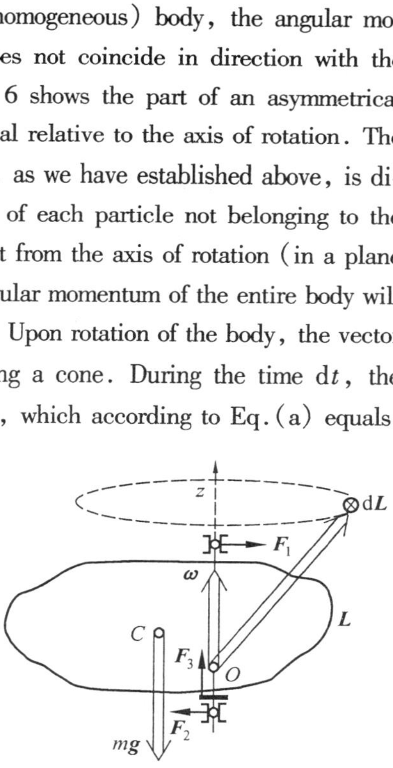


**Fig.3.7**

$$\mathrm{d}\boldsymbol{L} = \left(\sum \boldsymbol{M}_{ext}\right)\mathrm{d}t \tag{3.13}$$

If the vector $\boldsymbol{L}$ does not change in magnitude, then the vector $\mathrm{d}\boldsymbol{L}$ is directed beyond the drawing (Fig.3.7). The vector $\sum \boldsymbol{M}_{ext}$ has the same direction. In the example we are treating, the moments of the external forces include (1) the moment of the force of gravity mg directed toward us — we shall call it negative (this force is applied to the center of mass of the body $C$), (2) the positive moments of the forces, of lateral pressure of the bearings on the axis (the forces $\boldsymbol{F}_1$ and $\boldsymbol{F}_2$), and (3) the positive moment of the force of pressure of the bearing shoulder on the flange $\boldsymbol{F}_3$. We assume that friction forces are absent otherwise the vector $\boldsymbol{L}$ would not be constant in magnitude and $\mathrm{d}\boldsymbol{L}$ would not be perpendicular to $\boldsymbol{L}$.

The angular momentum relative to the axis of rotation for any (homogeneous or non-homogeneous, symmetrical or asymmetrical) body is

$$L_z = \sum L_{zi} = \sum m_i R_i^2 \omega = I\omega \tag{3.14}$$

[see Eqs (3.9) and (3.11)]. It must be stressed that unlike Eq. (3.12), Eq.3.14) is always correct.

Equation (a) states that

$$\frac{\mathrm{d}}{\mathrm{d}t} L_z = \sum M_{z,\,ext} \tag{b}$$

Introducing equation(b) into this expression Eq.(3.14) for $L_z$ we get

$$I\alpha_z = \sum M_{z,\,ext} \tag{3.15}$$

where $\alpha_z = \dot{\omega}$ is the projection of the angular acceleration onto the $z$-axis (we are considering rotation about a fixed axis. therefore the vector $\boldsymbol{\omega}$ can change only in magnitude). Equation (3.15) is similar to the equation $m\boldsymbol{a} = \boldsymbol{F}$. The part of the mass is played by the moment of inertia, that of the linear acceleration by the angular acceleration, and, finally, the part of the resultant force is played by the total moment of the external forces.

In the above example, the moments of all the external forces are perpendicular to the axis of rotation. Hence, their projections onto the $z$-axis equal zero. Accordingly, the angular velocity $\omega$ remains constant, which is what should be expected in the absence of friction.

We must point out that in the rotation of a homogeneous symmetrical body, forces of lateral pressure of the bearings on the axis (the forces $\boldsymbol{F}_1$ and $\boldsymbol{F}_2$ in Fig.3.7) do not appear. In the absence of the force of gravity, we could remove the bearings—the axis would retain its position in space without them. An axis whose position in space remains constant when bodies rotate about it in the absence of external forces is called a free axis of a body.

It is possible to prove that for a body of any shape and with an arbitrary arrangement of its mass there are three mutually perpendicular axes passing through the center of mass of the body that can be free axes. They are called the principal axes of inertia of the body.

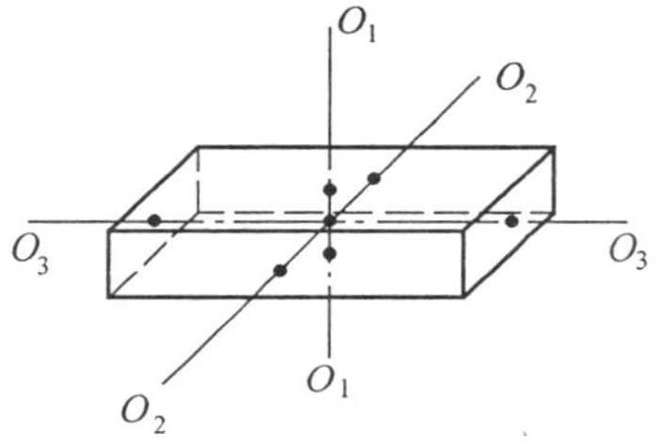


**Fig.3.8**

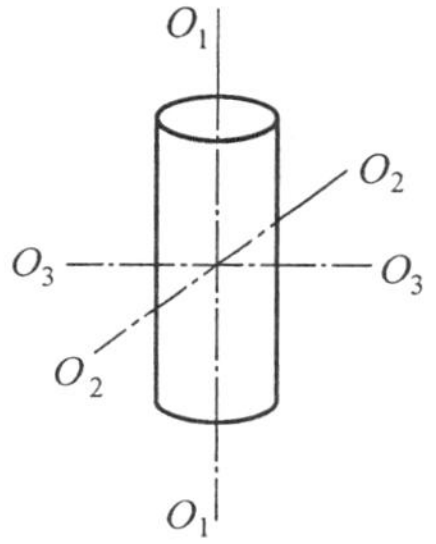


**Fig.3.9**

In a homogeneous parallelepiped (Fig.3.8), the principal axes of inertia are obviously the axes $O_1O_1$, $O_2O_2$ and $O_3O_3$ passing through the centers of opposite faces.

In bodies possessing axial symmetry (for example, in a homogeneous cylinder), the axis of symmetry is one of the principal axes of inertia. Any two mutually perpendicular axes in a plane at right angles to the axis of symmetry and passing through the center of mass of the body can be the other two principal axes (Fig.3.9). Thus, in such a body only one of the principal axes of inertia is fixed.

In a body with central symmetry, i.e. in a sphere whose density depends only on the distance from its center, any three mutually perpendicular axes passing through the center of mass are the principal axes of inertia. Consequently, none of the principal axes of inertia is fixed.

The moments of inertia relative to the principal axes are called the principal moments of inertia of a body. In the general case, those moments differ: $I_1 \neq I_2 \neq I_3$. For a body with axial symmetry, two of the principal moments of inertia are the same, while the third one, generally speaking, differs from them: $I_1 = I_2 \neq I_3$. And, finally, for a body with central symmetry, all three principal moments of inertia are the same $I_1 = I_2 = I_3$.

Not only a homogeneous sphere, but also, for instance, a homogeneous cube has equal values of the principal moments of inertia. In the general case, such equality may be observed for bodies of an absolutely arbitrary shape when their mass is properly distributed. All such bodies are called spherical tops. Their feature is that any axis passing through their center of mass has the properties of a free axis and, consequently, none of the principal axes is fixed, as for a sphere. All spherical tops behave the same when they rotate in identical conditions.

Bodies for which $I_1 = I_2 \neq I_3$ behave like homogeneous bodies of revolution. They are called symmetrical tops. Finally, bodies for which $I_1 \neq I_2 \neq I_3$ are called asymmetrical tops.

If a body rotates in conditions when there is no external action, then only rotation about the principal axes corresponding to the maximum and minimum values of the moment of inertia is stable. Rotation about an axis corresponding to an intermediate value of the moment will be unstable. This signifies that the forces appearing upon the slightest deviation of the axis of rotation from this principal axis act in a direction causing the magnitude of this deviation to grow. When the axis of rotation deviates from a stable axis, the forces produced return the body to rotation about the corresponding principal axis.

We can convince ourselves that what has been said above is true by tossing a body having the shape of a parallelepiped (for example, a match box) and simultaneously bringing it into rotation. We shall see that the body when falling can rotate stably about axes passing through the biggest or smallest faces. Attempts to toss the body so that it rotates about an axis passing through the faces of an intermediate size will be unsuccessful.

If an external force is exerted, for instance, by the string on which a rotating body is suspended, then only rotation about the principal axis corresponding to the maximum value of the moment of inertia will be stable. This is why a thin rod suspended by means of a string fastened to its end when brought into rapid rotation will in the long run rotate about an axis normal to it passing through its center (Fig.3.10a). A disk suspended by means of a string fastened to its edge (Fig.3.10b) behaves in a similar way.

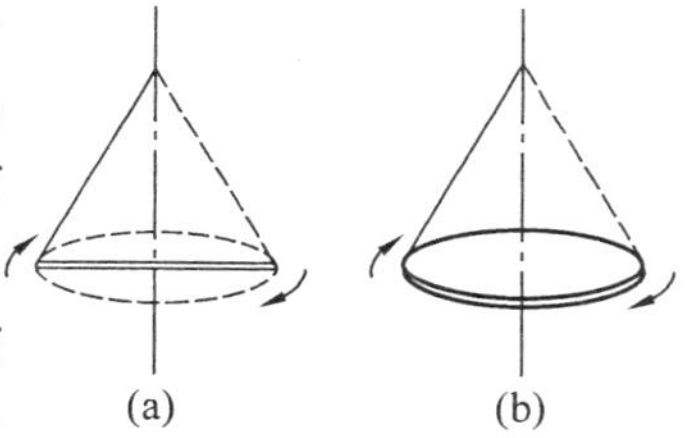

**Fig.3.10**

Up to now we have treated bodies with a constant distribution of their mass. Now let us assume that a rigid body can lose for a certain time its property of a constant arrangement of its parts, and within this time redistribution of the body's mass occurs that results in the moment of inertia changing from $I_1$ to $I_2$. If such a redistribution occurs in conditions when $\sum \boldsymbol{M}_{ext} = 0$, then in accordance with the law of conservation of angular momentum the following equation must be observed:

$$I_1\omega_1 = I_2\omega_2 \tag{3.16}$$

where $\omega_1$ is the initial, and $\omega_2$ is the final value of angular velocity of the body. Thus, a change in the moment of inertia leads to a corresponding change in the angular velocity. This explains why a spinning figure skater (or a man on a rotating platform) begins to rotate more slowly when he stretches his arms out, and gains speed when he presses his arms against his body.

## The Further Study:

### Moment of Inertia

From the definition of the moment of inertia

$$I = \sum \Delta m_i R_i^2$$

we can see that it is an additive quantity. This signifies that the moment of inertia of a body equals the sum of the moments of inertia of its parts.

We introduced the concept of the moment of inertia when dealing with the rotation of a rigid body. It must be borne in mind, however, that this quantity exists irrespective of rotation. Every body, regardless of whether it is rotating or at rest, has a definite moment of inertia relative to any axis, just like a body has a mass regardless of whether it is moving or at rest.

The distribution of the mass within a body can be characterized with the aid of a quantity called the density. If a body is homogeneous, i.e. its properties are the same at all of its points, then the density is defined as the quantity

$$\rho = \frac{m}{V} \tag{3.17}$$

where $m$ and $V$ are the mass and volume of the body, respectively. Thus, the density of a homogeneous body is the mass of a unit of its volume.

For a body with an unevenly distributed mass, Eq. (3.17) gives the average density. The density at a given point is determined in this case as follows:

$$\rho = \lim_{\Delta V \to 0} \frac{\Delta m}{\Delta V} = \frac{\mathrm{d}m}{\mathrm{d}V} \tag{3.18}$$

In this expression, $\Delta m$ is the mass contained in the volume $\Delta V$, which in the limit transition contracts to the point at which the density is being determined.

The limit transition in Eq. (3.18) must not be understood in the sense that $\Delta V$ contracts literally to a point. If such a meaning is implied, we would get a greatly differing result for two virtually coinciding points, one of which is at the nucleus of an atom, while the other is at a space between nuclei (the density for the first point would be enormous, and for the second one it would be zero). Therefore, $\Delta V$ should be diminished until we get an infinitely small volume from the physical viewpoint. We understand this to mean such a volume which on the one hand is small enough for the macroscopic (i.e. belonging to a great complex of atoms) properties within its limits to be considered identical, and on the other hand is sufficiently great to prevent discreteness (discontinuity) of the substance from manifesting itself.

By Eq. (3.18), the elementary mass $\Delta m_i$ equals the product of the density of a body $\rho_i$, at a given point and the corresponding elementary volume $\Delta V_i$:

$$\Delta m_i = \rho \Delta V_i$$

Consequently, the moment of inertia can be written in the form

$$I = \sum \rho_i R_i^2 \Delta V_i \tag{3.19}$$

If the density of a body is constant, it can be put outside the sum:

$$I = \rho \sum R_i^2 \Delta V_i \tag{3.20}$$

Equations (3.19) and (3.20) are approximate. Their accuracy grows with diminishing elementary volumes $\Delta V_i$ and the elementary masses $\Delta m_i$, corresponding to them. Hence, the task of finding the moments of inertia consists in integration:

$$I = \int R^2 \mathrm{d}m = \int \rho R^2 \mathrm{d}V \tag{3.21}$$

The integrals in Eq. (3.21) are taken over the entire volume of the body.

The quantities $\rho$ and $R$ in these integrals are position functions, i.e., for example, functions of the Cartesian coordinates $x$, $y$, and $z$.

As an example, let us find the moment of inertia of a homogeneous disk relative to an axis perpendicular to the plane of the disk and passing through its center (Fig. 3.11). Let us divide the disk into annular layers of thickness $dR$. All the points of one layer will be at the same distance $R$ from the axis. The volume of such a layer is

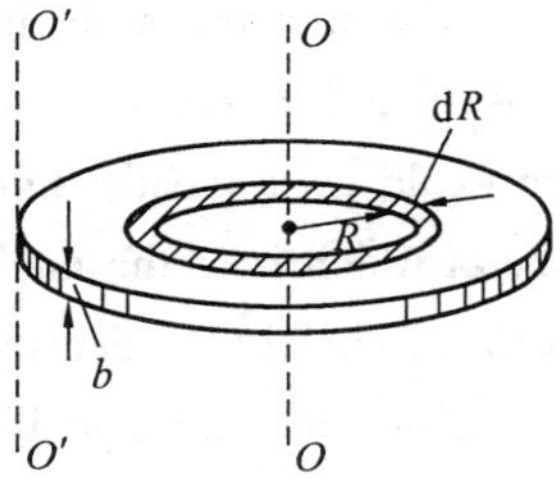

**Fig.3.11**

$$dV = b2\pi R dR$$

where $b$ is the thickness of the disk.

Since the disk is homogeneous, its density at all its points is the same, and $\rho$ in Eq.(3.21) can be put outside the integral:

$$I = \rho \int R^2 dV = \rho \int_0^{R_0} R^2 b 2\pi R dR$$

where $R_0$ is the radius of the disk. Let us put the constant factor $2\pi b$ outside the integral:

$$I = 2\pi b \rho \int_0^{R_0} R^3 dR = 2\pi b \rho \frac{R_0^4}{4}$$

Finally, introducing the mass of the disk $m$ equal to the product of the density $\rho$ and the volume of the disk $b\pi R_0^2$, we get

$$I = \frac{mR_0^2}{2} \qquad (3.22)$$

The finding of the moment of inertia in the above example was simplified quite considerably owing to the fact that the body was homogeneous and symmetrical, and we sought the moment of inertia relative to an axis of symmetry. If we wanted to find the moment of inertia of the disk relative, for example, to the axis $O'O'$ perpendicular to the disk and passing through its edge (see Fig. 3.11), the calculations would evidently be much more complicated. The finding of the moment of inertia is considerably simplified in such cases if we use the Steiner or parallel axis theorem, which is formulated as follows: the moment of inertia $I$ relative to an arbitrary axis equals the moment of inertia $I_C$ relative to an axis parallel to the given one and passing through the body's center of mass plus the product of the body's mass $m$ and the square of the distance $b$ between the axes:

$$I = I_C = mb^2 \tag{3.23}$$

According to the parallel axis theorem, the moment of inertia of the disk relative to the axis $O'O'$ equals the moment of inertia relative to the axis passing through the center of the disk, which we have found [Eq. (3.22)] plus $mR_0^2$ (the distance between the axes $O'O'$ and $OO$ equals the radius of the disk $R_0$):

$$I = \frac{mR_0^2}{2} + mR_0^2 = \frac{3}{2} mR_0^2$$

Thus, the parallel axis theorem in essence reduces the calculation of the moment of inertia relative to an arbitrary axis to the calculation of the moment of inertia relative to an axis passing through the center of mass of the body.

To prove the parallel axis theorem, let us consider axis $C$ passing through the center of mass of a body and axis $O$ parallel to it and at a distance $b$ from axis $C$ (Fig. 3.12, both axes are perpendicular to the plane of the drawing). Let $\boldsymbol{R}_i$ be a vector perpendicular to axis $C$ and drawn from the axis to the elementary mass $\Delta m_i$, and $\boldsymbol{R}_i$ be a similar vector drawn from axis $O$. We shall also introduce the vector $\boldsymbol{b}$ perpendicular to the axes and connecting the corresponding points of axes $O$ and $C$. For any pair of points opposite each

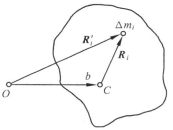

**Fig.3.12**

other, this vector has the same value (equal to the distance $b$ between the axes) and the same direction. The following relation holds between the vectors listed above:

$$\boldsymbol{R}'_i = \boldsymbol{b} + \boldsymbol{R}_i$$

The square of the distance to the elementary mass $\Delta m_i$ from axis $C$ is $R_i^2 = \boldsymbol{R}_i^2$, and from axis $O$ is

$$(\boldsymbol{R}'_i)^2 = (\boldsymbol{b} + \boldsymbol{R}_i)^2 = b^2 + 2\boldsymbol{b}\boldsymbol{R}_i + R_i^2$$

With a view to the above expression, the moment of inertia of the body relative to axis $O$ can be written in the form

$$I = \sum \Delta m_i R'^2_i = b^2 \sum \Delta m_i + 2\boldsymbol{b} \sum \Delta m_i \boldsymbol{R}_i + \sum \Delta m_i R_i^2 \tag{3.24}$$

(we have put the constant factors outside the sum). The last term in this expression is the moment of inertia of the body relative to axis $C$. Let us designate it $I_C$. The sum of the elementary masses gives the mass of the body m. The sum $\sum \Delta m_i \boldsymbol{R}_i$ equals the product of the mass of the body and the vector $\boldsymbol{R}$ drawn from axis $C$ to the center of mass of the body. Since the center of mass is on axis $C$, this vector $\boldsymbol{R}$ and, consequently, the second term in Eq. (3.24) vanish. We thus arrive at the conclusion that

$$I = mb^2 + I_C$$

Q.E.D. [see Eq. (3.23)].

In concluding, we shall give the values of the moments of inertia for selected bodies (the latter are assumed to be homogeneous, $m$ is the mass of the body)

1. The body is a thin long rod with a cross section of any shape. The maximum cross-sectional dimension $b$ of the rod is many times smaller than its length $l$ ($b \ll l$). The moment of inertia relative to an axis perpendicular to the rod and passing through its middle (Fig. 3.13) is

$$I = \frac{1}{12} ml^2 \tag{3.25}$$

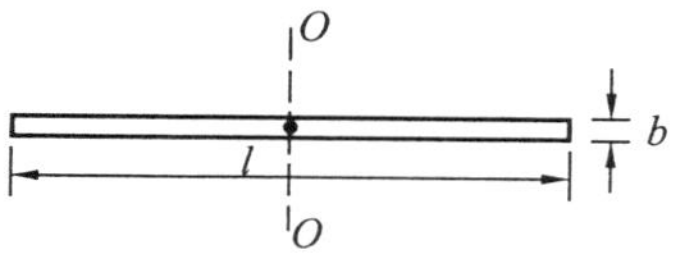


Fig.3.13

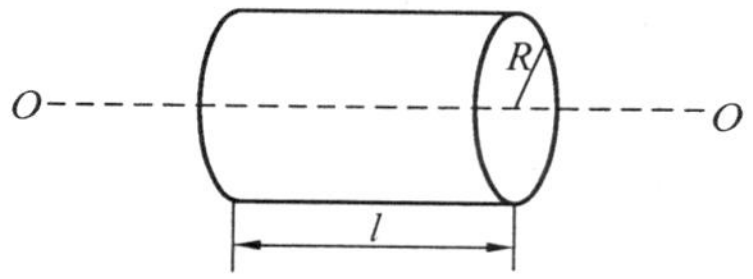


Fig.3.14

2. For a disk or cylinder with any ratio of $R$ to $l$ (Fig.3.14), the moment of inertia relative to an axis coinciding with the geometrical axis of the cylinder is

$$I = \frac{1}{2} mR^2 \tag{3.26}$$

3. The body is a thin disk. The thickness of the disk $b$ is many times smaller than the radius of the disk $R(b \ll R)$. The moment of inertia relative to an axis coinciding with the diameter of the disk (Fig.3.15) is

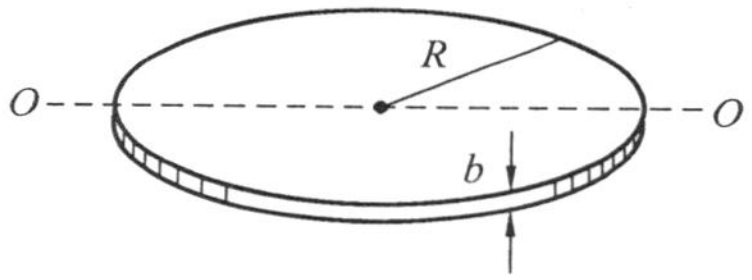


Fig.3.15

$$I = \frac{1}{4} mR^2 \tag{3.27}$$

4. The moment of inertia of a sphere of radius $R$ relative to an axis passing through its center is

$$I = \frac{2}{5} mR^2 \tag{3.28}$$

## Summary of Key Terms

**Rotational inertia** That property of an object to resist any change in its state of rotation: if at rest the body tends to remain at rest; if rotating, it tends to remain rotating and will continue to do so unless acted upon by a net external torque.

**Torque** The product of force and lever-arm distance, which tends to produce rotation.

**Center of mass** The average position of mass or the single point associated with an object where all its mass can be considered to be concentrated.

**Center of gravity** The average position of weight or the single point associated with an object where the force of gravity can be considered to act. Usually the same place as the center of mass.

**Equilibrium** The state of an object when not acted upon by a net force or net torque. An object in equilibrium may be at rest or moving at uniform velocity; that is not accelerating.

**Centripetal force** A center-seeking force that causes an object to follow a circular path.

**Angular momentum** A measure an object's rotation about a particular axis; more specifically the product of its rotational inertia and rotational velocity. For an object that is small compared to the radial distance, it is the product of mass, speed, and radial distance of rotation.

## Reading Materials:

### Blood Pressure

Taking one's blood pressure monitors the pumping action of the heart. When the heart contracts, the blood pressure in the arteries increases. When the heart relaxes, the pressure goes down. The "upper" or maximum pressure is called the systolic pressure, and the "lower" or minimum pressure is called the diastolic pressure.

When taking your blood pressure, a doctor or nurse (or even a coin – operated machine) is measuring the pressure of the blood on the arterial walls. This is done by instrument called a sphygmomanometer (Greek spbygmo, meaning "pulse"). This tongue twister is pronounced "ss – fig – mom – an – om – it – er". The dial pressure gauge is calibrated in mm

Hg or torr (Older types of meters use a mercury column to measure the pressure).

An inflatable cuff is placed snugly around the upper arm. When inflated to a sufficient pressure, the cuff shuts off the blood flow in the large arteries of the arm. Air is then slowly released from the cuff, and the doctor or nurse listens with a stethoscope over the artery just below the cuff in the bend of the arm.

When the cuff pressure is slightly lower than the blood pressure, blood begins to flow through the artery with each beat of the heart. The rhythmic escape of blood beneath the cuff produces a distinct sound that can be heard through the stethoscope. As soon as the sound is heard, the pressure is noted, for example, 120 mm Hg (systolic or "upper" pressure).

As more air escapes from the cuff, blood flows more steadily through the artery. At a specific lower pressure the distinct beating sound is no longer heard, for example, at 80 mm Hg (diastolic pressure). Blood pressure readings are commonly reported in a ratio form, such as [Trial mode] (a reading of "120 over 80"). Normally, the pressure in the arteries ranges between 100 and 140 systolic (contraction pressure) and between 70 and 90 diastolic (relaxation pressure).

## Scientist: Isaac Newton

On Christmas Day, 1642, Isaac Newton was born in Woolsthrope in Lincolnshire, England. His father had died three months earlier. When he was 14 years old, his mother became widowed for a second time and brought Isaac home from school to help run the family farm. He proved to be a lackadaisical farmer, being occupied more with mathematics than farm chores.

At the age of 18, he entered Trinity College at Cambridge and received his degree four years later in 1665. Later that year while he was preparing for advanced degree examinations, the spread of the Great Plague caused the university to close. Newton returned home, and during the next 18 months he conceived most of the ideas for his famous discoveries in science and mathematics. Chief among these were the development of calculus mathematics and studies on light and color, motion, and graviton. As Newton later described this period,

"I was in the prime of my age for invention, and minded Mathematics and Philosophy (science) more than at any time."

After the plague had passed, he returned to Cambridge and at the age of 26 was appointed professor of mathematics. Newton's early published works were in the field of optics. He developed a new type of telescope, a reflecting telescope, that used a mirror rather than a lens to collect light. His most notable book, principia mathematica Philosophiae Naturalis, or Principia for short, was published in 1687. The cost of the book was borne by a contemporary, Edmund Halley, who predicted the return of a famous comet that bears his name. In Principia Newton set forth his theories on gravity, tides, and motion.

A bachelor, Newton lived very austerely and is reported to have been the classic absent-minded professor, being so absorbed in his work that he forgot meals and other day-to-day activities. In later life Newton moved to London and was appointed master of the mint in 1701. Queen Anne knighted him in 1705 in recognition of his numerous accomplishments. Early in 1727 he became seriously ill and died on March 20 of that year.

An insight into the character of this great scientist is given in one of his statements:

"If I have been able to see further than some, it is because I have stood on the shoulders of giants."

# PART 2
# MOLECULAR PHYSICS

# 4

# General Information

## 4.1 STATISTICAL PHYSICS AND THERMODYNAMICS

Molecular physics is a branch of physics studying the structure and properties of a substance on the basis of the so-called molecular kinetic notions. According to these notions, any body—solid, liquid or gaseous—consists of an enormous number of exceedingly small separate particles-molecules. (Atoms can be considered as monatomic molecules.) The molecules of a substance are in disordered, chaotic motion having no predominating direction. Its intensity depends on the temperature of the substance.

A direct proof of the existence of chaotic motion of molecules is Brownian motion. This phenomenon consists in that very small (visible only in a microscope) particles suspended in a fluid are always in a state of continuous chaotic motion that does not depend on external causes and is a manifestation of the internal motion of the substance. The Brownian motion of particles is due to their chaotic collisions with molecules.

The object of the molecular-kinetic theory is to interpret the properties of bodies that are directly observed in experiments (pressure, temperature, etc.) as the summary result of the action of molecules. It uses the statistical method and is interested not in the motion of separate molecules, but only in average quantities characterizing the motion of an enormous combination of particles. This explains its other name-statistical physics.

Thermodynamics also studies various properties of bodies and changes in the state of a substance. Unlike the molecular-kinetic theory, however, thermodynamics studies macroscopic properties of bodies and natural phenomena without being interested in their microscopic picture. Thermodynamics permits us to arrive at a considerable number of conclusions on how processes go on without taking molecules and atoms into consideration and without treating the processes from a microscopic standpoint.

Thermodynamics is founded on several fundamental laws established as a result of generalizing a large amount of experimental facts. Consequently, the conclusions of thermodynamics have a very general nature.

By considering the changes in the state of a substance from different viewpoints, thermodynamics and the molecular-kinetic theory mutually supplements each other, forming in essence a single entirety.

Turning to the history of the development of molecular-kinetic notions, we must point out first of all that ideas on the atomistic structure of a substance were already advanced by the ancient Greeks. These ideas, however, were nothing more than a brilliant conjecture. In the 17th century, atomistics again came to the forefront, but as a scientific hypothesis instead of a conjecture. This hypothesis was developed especially greatly in the works of the outstanding Russian scientist Mikhail Lomonosov

(1711 ~ 1765). He attempted to give a single picture of all the physical and chemical phenomena known at his time. He proceeded from the corpuscular (according to modern terminology-molecular) notion of the structure of matter. Revolting against the theory of thermogen (a hypothetic thermal liquid whose content in a body determines the extent of its heating) that prevailed at his time, Lomonosov saw the "cause of heat" in the rotation of the particles of a body. Thus, Lomonosov in essence formulated molecular-kinetic ideas.

In the second half of the 19th century and at the beginning of the 20th century, atomistics became a scientific theory owing to the works of a number of scientists.

## 4.2 MASS AND SIZE OF MOLECULES

The masses of atoms and molecules are characterized by using quantities known as the **relative atomic mass of an element** (the atomic mass in short) and the **relative molecular mass of a substance** (the molecular mass). (These quantities were previously called the atomic weight and the molecular weight, respectively).

The atomic mass ($A_r$) of a chemical element is defined as the ratio of the mass of an atom of the element to 1/12 of the mass of the atom $^{12}C$ (This is the symbol for the carbon isotope with a mass number of 12). The molecular mass ($M_r$) of a substance is defined as the ratio of the mass of a molecule of the substance to 1/12 of the mass of the atom $^{12}C$. Their definitions show that the atomic and molecular masses are dimensionless Quantities.

A unit of mass equal to 1/12 of the mass of the atom $^{12}C$ is called the atomic mass unit(amu). Let us denote the value of this unit expressed in kilogrammes by the symbol $A_r m_{un}$. Hence, the mass of an atom expressed in kilogrammes will be mun and the mass of a molecule will be $M_r m_{un}$.

The amount of a substance containing a number of particles (atoms, molecules, ions, electrons, etc.) equal to the number of atoms in 0.012 kg of the carbon isotope $^{12}C$ is called a **mole** (the mole is a basic unit of the SI system). Multiple and submultiple units are also used such as the kilomole (kmol), the millimole (mmol) and the micromole ($\mu$mol).

The number of particles contained in a mole of a substance is called the **Avogadro constant**. It was found experimentally that the Avogadro constant is

$$N_A = 6.023 \times 10^{23}\ \text{mol}^{-1} \tag{4.1}$$

Thus, for example, a mole of copper contains $N_A$ atoms of copper, a mole of water contains $N_A$ molecules of water, a mole of electrons contains $N_A$ electrons, etc.

The mass of a mole is called the molar mass $M$. It is evident that $M$ equals the product of $N_A$ and the mass of a molecule $M_r m_{un}$.

$$M = N_A M_r m_{un} \tag{4.2}$$

For carbon $^{12}C$, we have $M = 0.012$ kg/mol, and the mass of an atom is $12 m_{un}$. Substitution of these values in Eq. (4.2) yields

$$0.012(\text{kg/mol}) = N_A(\text{mol}^{-1}) \times 12\ m_{un}(\text{kg})$$

Hence,

$$m_{un}(\text{kg}) = \frac{0.001(\text{kg}\cdot\text{mol}^{-1})}{N_A(\text{mol}^{-1})} = \frac{0.001}{6.023 \times 10^{23}} =$$

$$1.66 \times 10^{-27}\ \text{kg} = 1.66 \times 10^{-24}\ \text{g} \tag{4.3}$$

Hence, the mass of any atom is $1.66 \times 10^{-27} A_r$ kg, and the mass of any molecule is $1.66 \times 10^{-27} M_r$ kg.

It can be seen from Eq. (4.3) that the product $N_A m_{un}$ equals 0.001 kg/mol. Introducing this value in Eq. (4.2) we find that

$$M = 0.001 M_r\ \text{kg/mol} \tag{4.4}$$

or

$$M = M_r\ \text{g/mol} \tag{4.5}$$

Thus, the mass of a mole expressed in grammas numerically equals the relative molecular mass. It must be borne in mind, however, that whereas $M_r$ is a dimensionless quantity, $M$ has a dimension and is measured in kg/mol (or g/mol).

Now let us assess the size of molecules. It is natural to assume that molecules in a liquid are quite close to one another. We can therefore approximately find the volume of one molecule by dividing the volume of a mole of a liquid, for example, water, by the number of molecules in a mole $N_A$. One mole (i.e. 18 g) of water occupies a volume of 18 $cm^3$ = $18 \times 10^{-6}$ $m^3$. Hence, the volume falling to one molecule is

$$\frac{18 \times 10^{-6}}{6 \times 10^{23}} = 30 \times 10^{-30} \text{ m}^3$$

It follows that the linear dimensions of water molecules are approximately

$$\sqrt[3]{30 \times 10^{-30}} \approx 3 \times 10^{-10} \text{ m} = 3 \text{ Å}$$

The molecules of other substances also have dimensions of the order of a few angstroms. (The angstrom – Å – is a non-system unit of length equal to $10^{-10}$ m. It is very convenient in atomic physics.)

## 4.3 STATE A SYSTEM PROCESS

We shall call a combination of bodies being considered a system of bodies or simply a system. An example of a system is a liquid and the vapor in equilibrium with it. Particularly, a system may consist of one body.

Any system can be in different states distinguished by their temperature, pressure, volume, etc. Such quantities characterizing the state of a system are called **parameters of state.**

A parameter does not always have a definite value. If, for example, the temperature at different points of a body is not the same, then a definite value of the parameter $T$ cannot be ascribed to the body. In this case, the body is said to be in a **non-equilibrium state.** If such a body is isolated from other bodies and left alone, then its temperature will level out and take on the same value $T$ for all points – the body will pass over into an equilibrium state. This value of $T$ will not change until the body is brought out of its equilibrium state by external action.

The same may also occur with other parameters, for instance, with the pressure $p$. If we take a gas confined in a cylindrical vessel closed with a tightly fitted piston and begin to rapidly move the latter in, then a gas cushion will be formed under it in which the pressure will be greater than in the remaining volume of the gas. Consequently, the gas in this case cannot be characterized by a definite value of the pressure $p$, and its state will be a non-equilibrium one. If we stop the movement of the piston, however, then the pressure at different points of the volume will level out, and the gas will pass over to an equilibrium state.

The process of transition of a system from a non-equilibrium state to an equilibrium one is called a **relaxation process**, or simply **relaxation.** The time needed for such a transition is called the **relaxation time.** The relaxation time is defined as the time in which the initial deviation of a quantity from its equilibrium value diminishes $e$ times, where $e$ is the base of natural logarithms. Each parameter of a system has its own relaxation time. The greatest of these times plays the part of the relaxation time of the system.

Thus, by an **equilibrium** state of a system is meant a state in which all the parameters of the system have definite values remaining constant as long as is desired in unchanging external conditions.

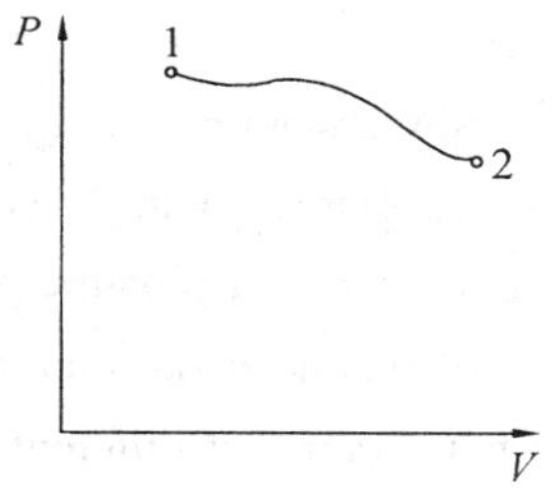

**Fig.4.1**

If we lay off the values of two parameters along coordinate axes then any equilibrium state of a system can be depicted by a point on the coordinate plane (see, point 1 in Fig.4.1). A no equilibrium state cannot be depicted in this way because at least one of the parameters will not have a definite value in this state.

**A process**, i.e. a transition of a system from one state to another, is associated with violation of the equilibrium of the system. Therefore, when a process occurs in a system, it passes through a sequence of non-equilibrium states. Reverting to the process of compressing a gas in a vessel closed with a piston that we have considered, we can conclude that the violation of equilibrium in moving in the piston is the greater, the faster the gas is compressed. If we move the piston in very slowly, equilibrium will be violated insignificantly, and the pressure at different points differs only slightly from a certain average value $p$. In the limit, if the gas is compressed infinitely slowly, it will be characterized at each moment by a definite value of the pressure. Consequently, the state of the gas at each moment in this case is equilibrium one, and the infinitely slow process will consist of a sequence of equilibrium states.

A process consisting of a continuous sequence of equilibrium states is called an **equilibrium** or **a quasistatic** one. It follows from what has been said above that only an infinitely slow process can be an equilibrium one. Real processes, when they occur sufficiently slowly, can approach an equilibrium one as close as **desired.**

An equilibrium process can be conducted in the reverse direction. The system will pass through the same states as in the forward process, but in the opposite sequence. This is why equilibrium processes are also called reversible.

A reversible (i.e. equilibrium) process can be depicted on a coordinate plane by the relevant curve (see Fig.4.1). We shall conditionally depict irreversible (i.e. non-equilibrium) processes by dash curves.

A process in which a system after a number of changes returns to its initial state is called a cyclic **process** or a **cycle.** The latter is depicted graphically by a closed curve.

The concepts of an equilibrium state and a reversible process play a great part in thermodynamics. All the quantitative conclusions of thermodynamics are strictly applicable only to equilibrium states and reversible processes.

## 4.4 INTERNAL ENERGY OF A SYSTEM

The internal energy of a body is defined as the energy of this body — the kinetic energy of the body as a whole and the potential energy of the body in the external force field. For example, in determining the internal energy of a mass of gas, we must not take into consideration the energy of motion of the gas together with the vessel containing it, and the energy due to the gas being in the field of the Earth's gravitational forces.

Hence, the concept of internal energy includes the kinetic energy of the chaotic motion of molecules, the potential energy of interaction between the molecules, and the intramolecular energy.

The internal energy of a system of bodies equals the sum of the internal energies of each of them separately and the energy of interaction between the bodies. The latter is the energy of intermolecular interaction in a thin layer on the interface between the bodies. This energy is so small in comparison with the energy of macroscopic bodies that it may be disregarded, and we may consider the internal energy of a system of macroscopic bodies as the sum of the internal energies of the bodies forming the system. The internal energy is thus an additive quantity.

The internal energy is a function of state of a system. This signifies that whenever a system is in a given state, its internal energy takes on the value characterizing this state regardless of the previous history of the system. Hence, the change in the internal energy when a system passes from one state to another will always equal the difference between the values of the internal energy in these states regardless of the path followed by the transition. In other words, the change in the internal energy does not depend on the process or processes that caused the system to pass from one state to another.

## 4.5 THE FIRST LAW OF THERMODYNAMICS

The internal energy can change in the main at the expense of two different processes: the performance of the work $A'$ on a body and the imparting of the heat $Q$ to it. The doing of work is attended by the displacement of the external bodies acting on the system. For example, when we move in the piston closing a vessel with a gas, the piston when moving does the work $A'$ on the gas. According to Newton's third law, the gas, in turn, does the work $A = -A'$ on the piston.

The imparting of heat to a gas is not associated with the motion of external bodies and is therefore not associated with the doing of macroscopic (i.e. relating to the entire complex of molecules which the body consists of) work on the gas. In this case, the change in the internal energy is due to the fact that separate molecules of the hotter body do work on separate molecules of the colder one. Energy is also transferred here by radiation. The combination of microscopic (i.e. involving not an entire body, but separate molecules of it) processes is called **heat transfer**.

Just as the amount of energy transferred by one body to another is determined by the work $A$ done by the bodies on each other, the amount of energy transmitted from one body to another by heat transfer is determined by the amount of heat $Q$ transferred by one body to the other. Thus, the increment of the internal energy of a system must equal the sum of the work $A'$ done on the system and the **amount of heat** $Q$ imparted to it:

$$U_2 - U_1 = Q + A' \tag{4.6}$$

Here $U_1$ and $U_2$ are the initial and final values of the internal energy of the system. It is customary practice to consider the work $A$ (equal to $A'$)

done by a system on external bodies instead of the work $A'$ done by external bodies on the system. Introducing $A$ in Eq. (4.6) instead of $A'$ and solving it relative to $Q$, we have

$$Q = U_2 - U_1 = A \tag{4.7}$$

Equation (4.7) expresses the law of energy conservation and forms the content of the **first law of thermodynamics**. It can be put in words as follows: the amount of heat imparted to a system is spent on an increment of the internal energy of the system and on the work done by the system on external bodies.

What has been said above does not at all signify that the internal energy of a system always grows when heat is imparted to it. It may happen that notwithstanding the transfer of heat to a system, its energy diminishes instead of growing ($U_2 < U_1$). In this case according to Eq. (4.7), we have $A > Q$, i.e. the system does work both at the expense of the heat $Q$ it has received and at the expense of its store of internal energy, whose decrement is $U_1 - U_2$. It must also be borne in mind that the quantities $Q$ and $A$ in Eq. (4.7) are algebraic ones ($Q < 0$ signifies that the system actually gives up heat instead of receiving it).

Examination of Eq. (4.7) shows that the amount of heat $Q$ can be measured in the same units as work or energy. In the SI system, the unit of the amount of heat is the joule.

A special unit called the **calorie** is also used to measure the amount of heat. One calorie equals the amount of heat needed to raise 1 g of water from 19.5 to 20.5℃. One kilocalorie equals 1 000 calories.

It was established experimentally that one – calorie is equivalent to 4.18 J. Hence; one joule is equivalent to 0.24 cal. The quantity $E = 4.18$ J/cal is called the **mechanical equivalent of heat**.

If the quantities in Eq. (4.7) are expressed in different units, then some of them must be multiplied by the appropriate equivalent. For example, if we express $Q$ in calories and $U$ and $A$ in joules, Eq. (4.7) must be written in the form

$$EQ = U_2 - U_1 + A$$

We shall always assume in the following that $Q$, $A$ and $U$ are expressed in the same units, and write the equation of the first law of thermodynamics in the form of Eq.(4.7).

In calculating the work done by a system or the heat received by it, we usually have to divide the process being considered into a number of elementary ones, each of which corresponds to a very small (infinitely small in the limit) change in the parameters of the system. Equation (4.7) has the following form for an elementary process:

$$\Delta' Q = \Delta U + \Delta' A \tag{4.8}$$

where $\Delta' Q$ = elementary amount of heat

$\Delta' A$ = elementary work

$\Delta U$ = increment of the internal energy of the system in the course of the given elementary process.

It is very important to bear in mind that $\Delta' Q$ and $\Delta' A$ must never be considered as increments of the quantities $Q$ and $A$. The change $\Delta$ in a quantity $f$ corresponding to an elementary process may be considered as the increment of this quantity only if $\sum \Delta f$ corresponding to the transition from one style to another does not depend on the path along which the transition occurs, i.e. if the quantity $f$ is a function of state. With respect to a function of state, we can speak of its "store" in each state. For example, we can speak of the store of internal energy that a system has in different states.

We shall see in the following that the quantity of work done by a system and the amount of heat it receives depend on the path followed by the system in its transition from one state to another. Hence, neither $Q$ nor $A$ are functions of state, and for this reason we cannot speak of the store of heat or work that a system has in different states.

Thus, the symbol $\Delta$ before $A$ and $Q$ is given a different meaning than that before $U$. To stress this circumstance, the $\Delta$ is primed in the former case. The symbol $\Delta U$ signifies an increment of the internal energy, whereas the symbols $\Delta' Q$ and $\Delta' A$ signify not an increment, but an elementary amount of heat and work.

To perform calculations, we pass over to differentials in Eq. (4.8). The equation of the first law thus acquires the following form:

$$\mathrm{d}'Q = \mathrm{d}U = \mathrm{d}'A \tag{4.9}$$

Integration of Eq. (4.9) over the entire process results in the expression

$$Q = (U_2 - U_1) + A$$

that is identical with Eq.(4.7).

We stress again that, for example, the result of integration of $\mathrm{d}'A$ must not be written in the form

$$\int_1^2 \mathrm{d}'A = A_2 - A_1$$

This form would mean that the work done by a system equals the difference between the values (i.e. the stores) of the work in the second and first states.

## 4.6 WORK DONE BY A BODY UPON CHANGES IN VOLUME

The interaction of a given body with bodies in contact with it can be characterized by the pressure, which it exerts on them. We can use pressure to describe the interaction of a gas with the walls of a vessel, and also of a solid or a liquid body with the medium (for example, a gas) surrounding it. The displacement of the points of application of the interaction forces is attended by a change in the volume of a body. Hence, the work done by a given body on external bodies can be expressed through the pressure and changes in the body's volume. Let us consider the following example to find this expression.

Assume that a gas is confined in a cylindrical vessel closed with a tightly fitting easily sliding piston (Fig.4.2). If for some reason or other the gas begins to expand, it will move the piston and do work on it. The elementary work done by the gas in moving the piston through the distance $\Delta h$ is

$$\Delta' A = F\Delta h$$

Where $F$ is the force with which the gas acts on the piston. Substituting for this force the product of the gas pressure $p$ and the piston area $S$, we have

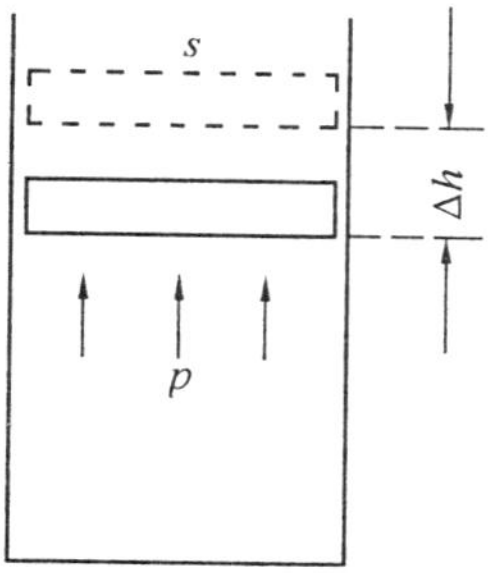


**Fig.4.2**

$$\Delta' A = pS\Delta h$$

But $S\Delta h$ is the increment of the volume of the gas $\Delta F$. Hence, the expression for the elementary work can be written as follows

$$\Delta' A = p\Delta V \qquad (4.10)$$

The quantity $\Delta' A$ in Eq.(4.10) is obviously an algebraic one. Indeed, in compression of the gas, the directions of the displacement $\Delta h$ and the force $F$ with which the gas acts on the piston are opposite. Consequently, the elementary work $\Delta' A$ will be negative. The increment of the volume $\Delta V$ in this case will also be negative. Thus, Eq (4.10) gives a correct expression for the work upon any changes in the volume of the gas.

If the pressure of the gas remains constant (for this to occur we must simultaneously change the temperature in the appropriate direction), the work done when the volume changes from $V_1$ to $V_2$ will be

$$A_{12} = P(V_2 - V_1) \qquad (4.11)$$

If a change in the volume is attended by a change in the pressure, then Eq (4.10) holds only for sufficiently small $\Delta V'S$. In this case the work done upon finite changes in the volume must be computed as the sum of elementary amounts of work expressed by Eq. (4.10), i.e. by integration:

$$A_{12} = \int_{v_1}^{v_2} p\mathrm{d}V \qquad (4.12)$$

The expressions found for the work hold for any changes in the volume of solid, liquid, and gaseous bodies. Let us consider another example to convince ourselves that this is true. Let us take a solid body of an arbitrary shape immersed in a liquid or gaseous medium that exerts on the body the pressure $p$ identical at all points (Fig. 4.3). Assume that the body expands so that separate elementary portions of its surface $\Delta S_i$ receive different displacements $\Delta h_i$. Hence, the $i$-th portion does the work $\Delta' A_i$ equal to $p\Delta S_i \Delta h_i$. The work done by the body can be found as the sum of the amounts of work done by separate portions:

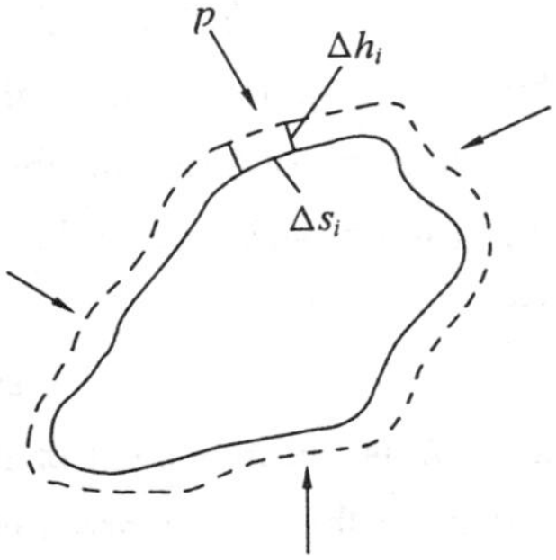

**Fig.4.3**

$$\Delta' S = \sum \Delta' A_i = \sum p\Delta S_i \Delta h_i$$

Factoring out of the sum the value of $p$ which is identical for all the portions and noting that $\sum \Delta S_i \Delta h_i$ gives the increment of the body's volume $\Delta V$, we can write that

$$\Delta' A = p\Delta V$$

i.e. in the general case we arrive at Eq. (4.10).

Let us depict the process of the change in the volume of the body in a $p - V$ diagram (Fig.4.4). The area of the narrow hatched strip in the diagram corresponds to the elementary work $\Delta' A_i = p_i \Delta V_i$. It is obvious that the area confined between the $V$-axis the curve $p = f(V)$, and the perpendiculars erected from points $V_1$ and $V_2$ numerically equals the work done when the volume changes from $V_1$ to $V_2$. The work done in a cyclic process numerically equals the area enclosed by the curve (Fig.4.5). It is obvious that the area confined between the $V$-axis the curve $p = f(V)$, and the perpendiculars erected from points $V_1$ and $V_2$ numerically equals the work done when the volume changes from $V_1$ to $V_2$. The work done in a cyclic process numerically equals the area enclosed by the curve (Fig. 4.5). Indeed, the work on path 1 – 2 is positive and numerically equals the area hatched diagonally to the right (we are considering a clockwise cycle). The work on path 2 – 1 is negative and numerically equals the area hatched diagonally to the left. Hence, the work during a cycle numerically equals the area enclosed by the curve. It will be positive in the direct cycle (i.e. in one conducted in the clockwise direction), and negative in the reverse cycle.

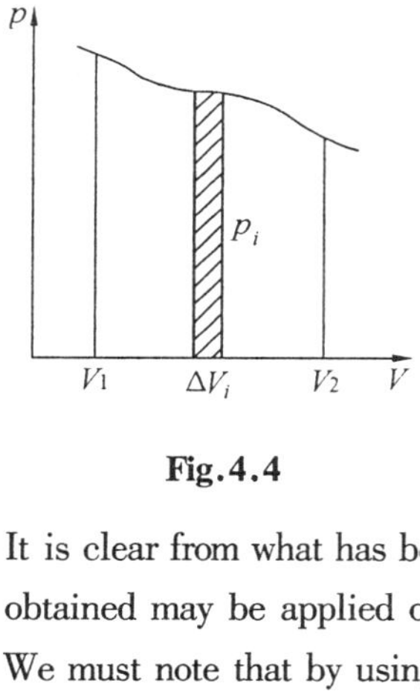


**Fig.4.4**

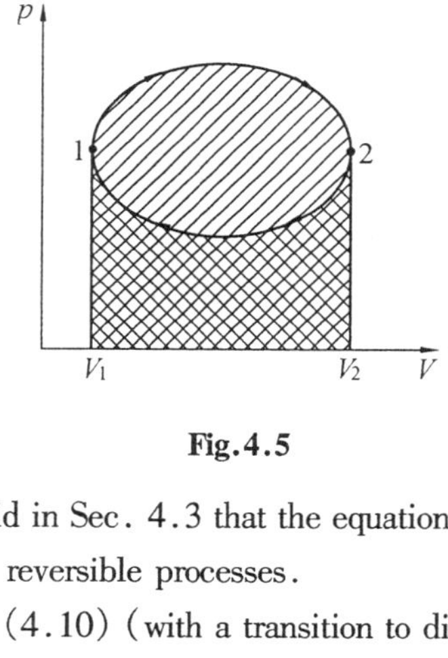


**Fig.4.5**

It is clear from what has been said in Sec. 4.3 that the equations we have obtained may be applied only to reversible processes.

We must note that by using Eq. (4.10) (with a transition to differentials), Eq. (4.9) expressing the first law of thermodynamics can be written as follows:

$$d'Q = dU = pdV \tag{4.13}$$

## 4.7 TEMPERATURE

We can arrive at a definition of the concept of temperature on the basis of the following reasoning. If contacting bodies are in a state of thermal equilibrium, i.e. do not exchange energy by heat transfer, they are said to have the same temperature. If when thermal contact is established between bodies one of them transmits energy to the other by heat transfer, then the first body is said to have a higher temperature than the second one. Many properties of bodies such as their volume and electrical resistance depend on the temperature. Any of these properties can be used for a quantitative definition of temperature.

Let us bring the body we have chosen for measuring the temperature (a thermometric body) into thermal equilibrium with melting ice. We shall assume that the body has a temperature of 0 centigrade and shall characterize quantitatively the property of the body (the temperature feature), which we intend to use for measuring the temperature. Let this feature be the volume of the body. Its value at 0 centigrade is $V_0$. Next we shall bring the same body into thermal equilibrium with water boiling under atmospheric pressure. Now we shall assume that the body in this state has a temperature of 100 centigrade, and shall determine the corresponding volume $V_{100}$. Presuming that the temperature feature we have chosen (the volume in the given example) changes linearly with the temperature, we can ascribe the following temperature to the state in which our thermometric body has the volume $V$:

$$t = \frac{V - V_0}{V_{100} - V_0} \times 100 \text{ degrees} \tag{4.14}$$

The temperature scale established in this way is called, as is known, the Celsius scale. An expression similar to Eq. (4.14) can also be written for the case when we use another temperature feature instead of the volume to measure the temperature.

After graduating a thermometer in this way, we can use it to measure the temperature by bringing it into thermal equilibrium with the body whose temperature we are interested in, and calculating the value of the volume.

When we compare thermometers functioning with different thermometric bodies (for example, mercury and alcohol) or different temperature features (for example, volume and electrical resistance) we find that the reading of these thermometers, which coincide at 0 and 100 centigrade owing to their being graduated at these temperatures, do not coincide at other temperatures. It thus follows that for the unique definition of a temperature scale, in addition to the way of graduation, we must also arrive at an agreement on the choice of the thermometric body and the temperature feature. How this choice is made in establishing the so – called empirical temperature scale will be treated in the following section. Getting ahead, we shall indicate that the second law of thermodynamics can serve as the basis of a temperature scale not depending on the properties of the thermometric body. This scale is called the **thermodynamic temperature scale.**

The **international practical temperature** scale of 1968, formerly called the Celsius (centigrade) scale is used in engineering and for everyday purposes. Physicists find the absolute scale more convenient. The temperature $T$ measured according to this scale is related to the temperature $t$ according to the Celsius scale by the equation

$$T = t + 273.15$$

The unit of absolute temperature is the kelvin (K). It was previously called the degree Kelvin (K). The international practical temperature is measured in degrees Celsius (℃). The sizes of the kelvin and the degree Celsius are the same. A temperature of 0 K is referred to as absolute zero, and $t = -273.15$℃ corresponds to it.

In the following, we shall show that the absolute temperature is proportional to the mean kinetic energy of translational motion of the molecules of a substance. This is the physical meaning of absolute temperature.

## 4.8 EQUATION OF STATE OF AN IDEAL GAS

The state of a given mass of a gas is determined by the values of three parameters: the pressure $p$, volume $V$, and temperature $T$ these parameters are related to one another according to a definite law so that a change in one of them causes a change in the others. This nation can be given analytically in the form of the function

$$F(p, V, T) = 0 \tag{4.15}$$

An expression determining the relation between the parameters of a body is called an equation of state of the body. Hence Eq. (4.15) is an equation of state of a given mass of a gas.

A gas, the interaction between whose molecules is negligibly small, has the simplest properties. Such a gas is called **ideal** (or perfect). The interaction between the molecules of any gas becomes negligibly small at a great rarefaction, i.e. at low densities of the gas. A real gas upon sufficient rarefaction is close in its properties to an ideal one. Some gases such as air, nitrogen, and oxygen differ only slightly from an ideal gas even in usual conditions, i.e. at room temperature and atmospheric pressure. Helium and hydrogen are especially close to an ideal gas in their properties.

Gases at low densities obey the following equation with a good accuracy:

$$\frac{pV}{T} = \text{const} \tag{4.16}$$

Consequently, this equation is **an equation of state of an ideal gas.**

According to the law established by Amadeo Avogadro (1776 ~ 1856), the moles of all gases occupy an identical volume in identical conditions (i.e. at the same temperature and pressure). In particular in the so-called standard conditions, i.e. at 0℃ and a pressure of 1 atm ($1.01 \times 10^5$ Pa), the volume of amole of any gas is $\text{dm}^3/\text{mol} = 22.4 \times$

$10^{-3}$ $m^3/mol$. It thus follows that when the amount of a gas is one mole, the value of the constant in Eq. (4.16) will be the same for all gases. Denoting the value of this constant corresponding to one mole by the symbol $R$, we can write Eq. (4.16) as follows:

$$pV_m = RT \tag{4.17}$$

We have used the subscript "$m$" with $V$ to show that we are dealing with the volume occupied by one mole of a gas at the given $p$ and $T$. Equation (4.17) is an equation of state of an ideal gas written for one mole.

The quantity $R$ is called the molar gas constant. According to Eq. (4.17) and Avogadro's law,

$$R = \frac{pV_m}{T} = \frac{1.01\times10^5\times22.4\times10^{-3}\ \mathrm{Pa\cdot m^3/mol}}{\mathrm{K}} = 8.31\ \frac{\mathrm{J}}{\mathrm{mol\cdot K}}$$

For practical calculations, it is sometimes convenient to use $R$ expressed in litre – atmospheres per mole kelvin:

$$R = \frac{1\ \mathrm{atm}\cdot22.4\ \mathrm{l/mol}}{273\ \mathrm{K}} = 0.0820\ \frac{\mathrm{l\cdot atm}}{\mathrm{mol\cdot K}}$$

It is a simple matter to pass over from Eq. (4.17) for one mole to an equation for any mass $m$, taking into account that at the same pressure and temperature $\nu$ moles of a gas will occupy a volume $\nu$ times greater than that occupied by one mole: $V = \nu V_m$. Multiplying Eq. (4.17) by $\nu = m/M$ (here $m$ is the mass of the gas and $M$ the mass of a mole) and introducing $V$ instead of $\nu V_M$, we get

$$pV = \frac{m}{M}RT \tag{4.18}$$

This equation is an equation of state of an ideal gas written for the mass $m$ of a gas.

Equation (4.18) can be given a different form. For this purpose we shall introduce the quantity

$$k = \frac{R}{N_A} \tag{4.19}$$

($R$ is the molar gas constant, and $N_A$ is the Avogadro constant). This quantity is known as the **Boltzmann constant.** It has a deeper physical meaning than the constant $R$. We shall show that $k$ is the constant of proportionality between the mean energy of thermal motion of a molecule and the absolute temperature substitution of the numerical values for $R$ and $N_A$ in Eq. (4.19) yields

$$k = \frac{8.31 \text{ J/(mol}\cdot\text{K)}}{6.023 \times 10^{23} \text{ mol}^{-1}} = 1.38 \times 10^{-23} \text{ J/K}$$

Let us multiply and divide the right-hand side of Eq. (4.18) by $N_A$. The equation can therefore be written in the form

$$pV = \nu N_A kT$$

The product ($\nu N_A$) equals the number of molecules $N$ contained in the mass m of a gas. Taking this into consideration, we find that

$$pV = NkT \tag{4.20}$$

Now let us divide both sides of Eq. (4.20) by $V$. Since $N/V = n$ is the number of molecules in a unit volume, we arrive at the equation

$$p = nkT \tag{4.21}$$

Equations (4.18), (4.20), and (4.21) are different forms of writing the equation of state for an ideal gas.

The ratio of the mass of a gas to the volume it occupies gives the density of the gas $\rho = m/V$. According to Eq. (4.18), the density of an ideal gas is determined by the expression

$$\rho = \frac{Mp}{RT} \tag{4.22}$$

Thus, the density of an ideal gas is proportional to the pressure and inversely proportional to the temperature.

The simple relation between the temperature and the remaining parameters of an ideal gas makes it tempting to use it as a thermometric substance. Ensuring a constant volume and using the pressure of the gas as the temperature feature, we can obtain a thermometer with an ideally linear temperature scale. In the following, we shall call this scale the ideal gas temperature scale.

In practice, according to an international agreement, hydrogen is taken as the thermometric body. The scale established for hydrogen with the use of Eq. (4.18) is called the **empirical temperature scale.**

## The Further Study:

### Internal Energy and Heat Capacity of an Ideal Gas

Experiments show that the internal energy of an ideal gas depends only on the temperature:

$$U = BT \tag{4.23}$$

Here $B$ is a coefficient of proportionality that remains constant within quite a broad range of temperatures.

The failure of the internal energy to depend on the volume occupied by a gas indicates that the molecules of an ideal gas do not interact with one another the overwhelming part of the time. Indeed, if the molecules did interact with one another, the internal energy would contain as an addend the potential energy of interaction, and the latter would depend on the mean distance between the molecules, i.e. on $V^{1/3}$.

It must be noted that interaction should take place upon collision, i.e. when the molecules come very close to one another; such collisions are very few in number in a rarefied gas, however. Each molecule spends the predominating part of its time in free night.

The heat capacity of a body is defined as the quantity equal to the amount of heat that must be imparted to the body to raise its temperature by one kelvin. If the amount of heat $d'Q$ imparted to a body raises its temperature by $dT$, then its heat capacity by definition is

$$C = \frac{d'Q}{dT} \tag{4.24}$$

This quantity is measured in joules per kelvin (J/K).

We shall denote the capacity of a mole of a substance, called the molar heat capacity, by the symbol $C$. It is measured in joules per mole kelvin [J/(mol)·K]

The heat capacity of a unit mass of a substance is called the specific heat capacity. We shall use the symbol c for it. The quantity c is measured in joules per kilogramme-kelvin [J/(kg)·K].

The following relation obviously holds between the heat capacity of a mole of a substance and the specific heat capacity of the same substance:

$$c = \frac{C}{M} \tag{4.25}$$

($M$ is the molar mass).

The value of the heat capacity depends on the conditions in which a body is heated. The heat capacity for heating at a constant volume or a constant pressure is of the greatest interest. The heat capacities at constant volume and constant pressure are designated by $C_V$ and $C_p$, respectively.

When heating occurs at constant volume, a body does no work on external bodies, and, consequently, according to the first law of thermodynamics, all the heat is spent on the increment of the internal energy of the body:

$$\mathrm{d}'Q_V = \mathrm{d}U \tag{4.26}$$

It can be seen from Eq. (4.26) that the heat capacity of any body at constant volume is

$$C_V = \left(\frac{\partial U}{\partial T}\right)_V \tag{4.27}$$

This notation stresses the fact that when differentiating the expression for $U$ with respect to $T$, the volume must be considered constant. For an ideal gas, $U$ depends only on $T$, and Eq. (4.27) can be written in the form

$$C_V = \frac{\mathrm{d}U_{\mathrm{m}}}{\mathrm{d}T}$$

(to obtain the molar heat capacity of a gas, we must take the internal energy of a mole).

Equation (4.23) for one mole of a gas has the form $U_m = B_m T$. Differentiating it with respect to $T$, we find that $C_V = B_m$. Thus, the expression for the internal energy of one mole of an ideal gas can be written in the form

$$U_m = C_V T \tag{4.28}$$

Where $C_V$ is a constant quantity — the molar heat capacity of a gas at constant volume.

The internal energy of an arbitrary mass $m$ of a gas will equal the internal energy of one mole multiplied by the number of moles of the eras in the mass $m$:

$$U = \frac{m}{M} C_V T \tag{4.29}$$

If a gas is heated at constant pressure, it will expand, doing positive work on external bodies. Consequently, more heat will be needed to raise the temperature of the gas by one kelvin in this case than when heating it at constant volume—part of the heat will be used by the gas to do work. Hence, the heat capacity at constant pressure must be greater than that at constant volume.

Let us write Eq. (4.13) of the first law of thermodynamics for a mole of a gas:

$$d'Q_p = dU_m + p\,dV_m \tag{4.30}$$

The subscript $p$ of $d'Q$ in this expression indicates that heat is imparted to the gas in conditions when $p$ is constant. Dividing Eq. (4.30) by $dT$, we get an expression for the molar heat capacity of a gas at constant pressure:

$$C_p = \frac{dU_m}{dT} + p\left(\frac{\partial V_m}{\partial T}\right)_p \tag{4.31}$$

The addend $dU_m/dT$ equals, as we have seen, the molar heat capacity of a gas at constant volume. Therefore, Eq. (4.31) can be written as follows:

$$C_p = C_V + p(\frac{\partial V_m}{\partial T})_p \tag{4.32}$$

The quantity $(\frac{\partial V_m}{\partial T})_p$ is the increment of the volume of a mole when the temperature is raised by one kelvin obtained with $p$ being constant. According to the equation of state (4.17), we have $V_m = RT/p$. Differentiating this expression with respect to $T$ provided that $p$ = const, we find

$$(\frac{\partial V_m}{\partial T})_p = \frac{R}{p}$$

Finally, using this result in Eq. (4.32), we get

$$C_p = C_v + R \tag{4.33}$$

Thus, the work done by a mole of an ideal gas when its temperature is raised by one kelvin at constant pressure equals the molar gas constant. It must be noted that Eq. (4.33) has been obtained by using an equation of state for an ideal gas and, consequently, holds only for an ideal gas.

The quantity

$$\gamma = \frac{C_p}{C_v} \tag{4.34}$$

is a quantity characterizing every gas. For monatomic gases, its value is close to 1.67, for biatomic gases to 1.4, for triatomic gases to 1.33, etc. In the following, we shall see that the value of $\gamma$ is determined by the number and the nature of the degrees of freedom of the molecule.

Substituting for $C_p$ in Eq. (4.34) its value from Eq. (4.33), we have

$$\gamma = \frac{C_V + R}{C_V} = 1 + \frac{R}{C_V}$$

Whence

$$C_V = \frac{R}{\gamma - 1} \tag{4.35}$$

Using this value of $C_V$ in Eq. (4.29), we get the following expression:

$$U = \frac{m}{M}\frac{RT}{\gamma - 1} \tag{4.36}$$

Comparison with Eq. (4.18) gives still another expression for the internal energy of an ideal gas:

$$U = \frac{1}{\gamma - 1}pV \tag{4.37}$$

## Summary of Key Terms

**Temperature** A measure of the average kinetic energy per molecule in a substance, measure in degrees Celsius or Fahrenheit or in kelvins.

**Absolute zero** The lowest possible temperature that a substance may have – the temperature at which molecules of a substance have their minimum kinetic energy.

**Heat** The energy that flows from a substance of higher temperature to a substance of lower temperature, commonly measured in calories or joules.

**Internal energy** The total of all molecular energies, kinetic plus potential energy, internal to a substance.

**Specific heat capacity** The quantity of heat per unit mass required to raise the temperature of a substance by 1 Celsius degree.

# 5

# Statistical Physics

## 5.1 INFORMATION FROM THE THEORY OF PROBABILITY

Assume that we have a macroscopic system, i.e. a system formed by an enormous number of microparticles (molecules, atoms, ions, electrons), in a given state. Assume further that a quantity $x$ characteristic of the system can have the discrete values

$$x_1, x_2, x_3, \cdots, x_i, \cdots, x_k, \cdots, x_s$$

Let us make a very great number $N$ of measurements of the quantity $x$, bringing the system before each measurement to the same initial state. Instead of performing repeated measurements of the same system, we can take $N$ identical systems in the same state and measure the quantity $N$ once in all these systems. Such a set of identical systems in an identical state is called a statistical ensemble.

Assume that $N_1$ measurements gave the result $x_1$, $N_2$ measurements the result $x_2, \cdots, N_i$ measurements the result $x_1$, and so on ( $\sum N_i = N$ is the number of systems in the ensemble). The quantity $N_i/N$ is defined as the relative frequency of appearance of the result $x_i$ while the limit of this quantity obtained when $N$ tends to infinity, i.e.

$$P_i = \lim_{N\to\infty} \frac{N_i}{N} \tag{5.1}$$

is called the probability of appearance of the result $x_i$. In the following, in order to simplify the equations, we shall write the expression for the probability in the form $N_i/N$, bearing in mind that the transition to the limit is performed at $N\to\infty$.

Since $\sum N_i = N$, we have

$$\sum P_i = \sum \frac{N_i}{N} = 1 \tag{5.2}$$

i.e. the sum of the probabilities of all possible results of measurement equals unity.

The probability of obtaining the result $x_i$ or $x_k$ is

$$P_{i \text{ or } k} = \frac{N_i + N_k}{N} = \frac{N_i}{N} + \frac{N_k}{N} = P_i + P_k$$

We have thus arrived at the theorem of summation of probabilities. It states that

$$P_{i \text{ or } k} = P_i + P_k \tag{5.3}$$

Assume that a system is characterized by the values of two quantities $x$ and $y$. Both quantities can take on discrete values whose probabilities of appearance are

$$P(x_i) = \frac{N(x_i)}{N}, P(y_k) = \frac{N(y_k)}{N}$$

Let us find the probability $P(x_i, y_k)$ of the fact that a certain measurement will give the result $x_i$ for $x$ and $y_k$ for $y$. The result $x_i$ is obtained in a number of measurements equal to $N(x_i) = P(x_i)N$. If the value of the quantity $y$ does not depend on that of $x$, then the result $y_k$ will be obtained simultaneously with $x_i$ in a number of cases equal to

$$N(x_i, y_k) = N(x_i)P(y_k) = [P(x_i)N]P(y_k)$$

[$N(x_i)$ plays the part of $N$ for $y$]. The required probability is

$$P(x_i, y_k) = \frac{N(x_i, y_k)}{N} = P(x_i)P(y_k)$$

Now we have arrived at the theorem of multiplication of probabilities according to which the probability of the simultaneous occurrence of statistically independent events equals the product of the probabilities of each of them occurring separately:

$$P(x_i, y_k) = P(x_i)P(y_k) \tag{5.4}$$

Knowing the probability of the appearance of different measurement results, we can find the mean value of all the results. According to the definition of the mean value

$$\langle x \rangle = \frac{\sum N_i x_i}{N} = \sum P_i x_i \tag{5.5}$$

Let us extend the results obtained to the case when the quantity $x$ characterizing a system can take on a continuous series of values from zero to infinity. In this case, the quantity $x$ is said to have a continuous spectrum of values (in the previous case the spectrum of values was discrete).

Let us take a very small quantity $a$ (say, $a = 10^{-6}$) and find the number of measurements $\Delta N_0$ which give $0 < x < a$, the number $\Delta N_1$ which give $a < x < 2a, \cdots$, the number $\Delta N_x$ for which the result of the measurements is within the interval from $x$ to $x + a$, and so on. The probability of the fact that the result of the measurements will be within the interval from zero to a is $\Delta P_0 = \Delta N_0 / N$, within the interval from a to 2a is $\Delta P_1 = \Delta N_1 / N, \cdots$, within the interval from $x$ to $x + a$ is $\Delta P_x = \Delta N_x / N$. Let us draw an $x$-axis and lay off strips of width a and of height $\Delta P_x / a$ upward from it (Fig. 5.1a). We obtain a bar graph or histogram. The area of the bar whose left-hand edge has the coordinate $x$ is $\Delta P_x$, and the area of the entire histogram is unity [see Eq. (5.2)].

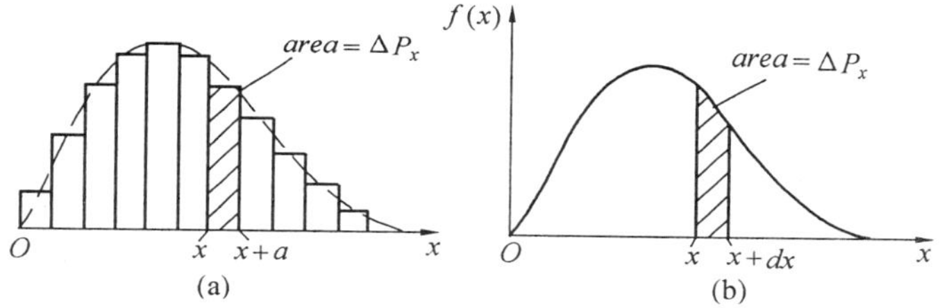


**Fig.5.1**

A histogram characterizes graphically the probability of obtaining results of measurements confined within different intervals of width $a$. The smaller the width of the interval $a$, the more detailed will the distribution of the probabilities of obtaining definite values of $x$ be characterized. In the limit when $a \to 0$, the stepped line confining the histogram transforms into a smooth curve (Fig. 5.1b). The function $f(x)$ defining this curve analytically is called a probability distribution function.

In accordance with the procedure followed in plotting the distribution curve, the area of the bar of width $\mathrm{d}x$ (see Fig. 5.1b) equals the probability of the fact that the result of a measurement will be within the range from $x$ to $x + \mathrm{d}x$. Denoting this probability by $\mathrm{d}P_x$, we can write that

$$\mathrm{d}P_x = f(x)\mathrm{d}x \tag{5.6}$$

The subscript "$x$" used with $\mathrm{d}p$ indicates that we have in mind the probability for the interval whose left-hand edge is at the point with the coordinate $x$. The area confined by a distribution curve, like that of a histogram, equals unity. This signifies that

$$\int f(x)\mathrm{d}x = \int \mathrm{d}P_x = 1 \tag{5.7}$$

Integration is performed over the entire interval of possible values of the quantity $x$. Equation (5.7) is an analogue of Eq. (5.2).

Knowing the distribution function $f(x)$, we can find the mean value of the result of measuring the quantity $x$. In $\mathrm{d}N_x = N\mathrm{d}P_x$ cases, a result equal to $x$ is obtained. The sum of such results is determined by the ex-

*pression* $x\mathrm{d}N_x = xN\mathrm{d}P_x$. The sum of all the possible results is $\int x\mathrm{d}N_x = \int xN\mathrm{d}P_x$. Dividing this sum by the number of measurements $N$, we get the mean value of the quantity $x$:

$$\langle x \rangle = \int x\mathrm{d}P_x \tag{5.8}$$

This equation is an analogue of Eq. (5.5).

Using Eq. (5.6) for $\mathrm{d}P_x$ in Eq. (5.8), we obtain

$$\langle x \rangle = \int xf(x)\mathrm{d}x \tag{5.9}$$

Similar reasoning shows that the the mean value of a function $\varphi(x)$ can be calculated by the equation

$$\langle \varphi(x) \rangle = \int \varphi(x)f(x)\mathrm{d}x \tag{5.10}$$

For examqle,

$$\langle x^2 \rangle = \int x^2 f(x)\mathrm{d}x \tag{5.11}$$

## 5.2 NATURE OF THE THERMAL MOTION OF MOLECULES

If a gas is in equilibrium, its molecules move absolutely without order, chaotically. All the directions of motion are equally probable, and none of them can be given preference over others. The velocities of the molecules may have the most diverse values. Upon each collision with other molecules, the magnitude of the velocity or speed of a given molecule should, generally speaking, change. It may grow or diminish with equal probability.

The velocities of molecules change by chance upon collisions. A molecule in a series of consecutive collisions may receive energy from its collision partners, and as a result its energy will considerably exceed the mean value $\langle \varepsilon \rangle$. Even if we imagine the absolutely fantastic case, however, in which all the molecules of a gas give up their energy to a single molecule and stop moving, the energy of this molecule, and consequently its velocity too, will still be finite. Thus, the velocity of molecules of a gas cannot have values beginning with a certain $v_{max}$ and ending with infinity. Taking into consideration that processes which would lead to the concentration of a considerable portion of the total energy of all the molecules on one molecule have a low probability, we can say that very high velocities in comparison with the mean value of the velocity can be realized extremely rarely. In exactly the same way, it is virtually impossible for the velocity of a molecule to vanish completely as a result of collisions. Hence, very low and very high velocities in comparison with the mean value have a low probability. The probability of the given value of $v$ tends to zero both when $v$ tends to zero and when it tends to infinity. It thus follows that the velocities of molecules are mainly grouped near a certain most probable value.

The chaotic nature of motion of molecules can be illustrated with the aid of the following procedure. Let us surround point $O$ with a sphere of arbitrary radius $r$ (Fig. 5.2). Any point $A$ on this sphere determines the direction from $O$ to $A$. Consequently, the direction in which the molecules of a gas move at a certain moment can be set by points on the sphere. The equal probability of all the directions results in the fact that the points showing the directions of motion of the molecules will be distributed over the sphere with a constant density. The latter equals the number $N$ of molecules being considered divided by the surface area of the sphere $4\pi r^2$. Collisions lead to changes in the directions of motion of the molecules. As a result, the positions of the $N$ points on the sphere continuously change. Owing to the chaotic nature of the motion of the molecules, however, the density of the points at any spot on the sphere

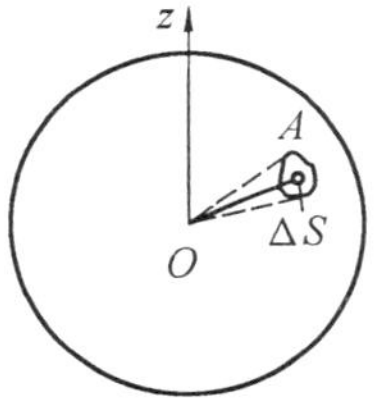


**Fig.5.2**

remains constant all the time.

The number of possible directions in space is infinitely great. But at each moment a finite number of directions is realized, equal to the number of molecules being considered. Therefore, putting the question of the number of molecules having a given (depicted by the point on the sphere) direction of motion is deprived of all meaning. Indeed, since the number of possible directions is infinitely great, whereas the number of molecules is finite, the probability of at least one molecule flying in a strictly definite direction equals zero. A question we are able to answer is what number of molecules move in directions close to the given one (determined by point $A$ on the sphere). All the points of the surface elements $\Delta S$ of the sphere taken in the vicinity of point $A$ (see Fig.5.2) correspond to these directions. Since the points depicting the directions of motion of the molecules are distributed uniformly over the sphere, then the number of points within the area $\Delta S$ will be

$$\Delta N_{\mathrm{A}} = N \frac{\Delta S}{4\pi r^2} \tag{5.12}$$

The subscript $A$ indicates that we have in view the molecules whose directions of motion are close to that determined by point $A$.

The ratio $\Delta S / r^2$ is the solid angle $\Delta \Omega$ subtended by the area $\Delta S$. Therefore, Eq. (5.12) can be written as follows:

$$\Delta N_{\mathrm{A}} = N \frac{\Delta \Omega}{4\pi} \tag{5.13}$$

Here $\Delta \Omega$ is the solid angle containing the directions of motion of the molecules being considered. We remind our reader that $4\pi$ is a complete solid angle (corresponding to the entire surface of the sphere).

The direction of $OA$ can be given with the aid of the polar angle $\theta$ and the azimuth $\varphi$ (Fig. 5.3). Hence, the directions of motion of the molecules of a gas can be characterized by giving for each molecule the values of the angles $\theta$ and $\varphi$ measured from a fixed direction $OZ$ (we can take the direction of a normal to the surface of the vessel confining a gas as such a direction) and the plane $P_o$, drawn through it.

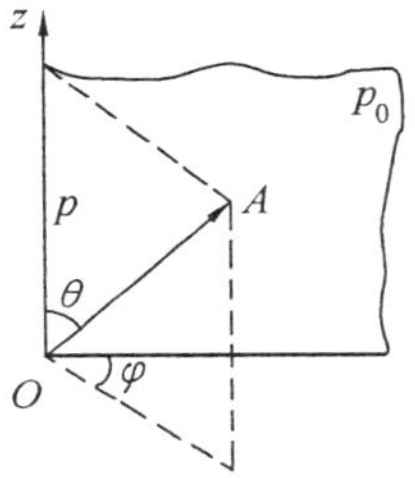


**Fig.5.3**

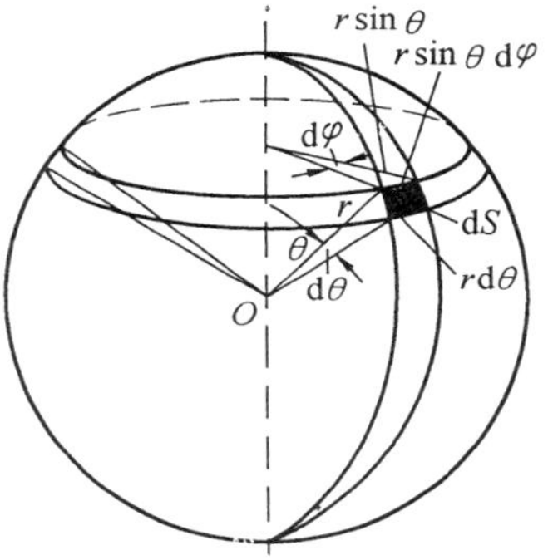


**Fig.5.4**

Let us surround the origin of coordinates $O$ with a sphere of radius r and find the element $\mathrm{d}S$ of the sphere corresponding to the increments $\mathrm{d}\theta$ and $\mathrm{d}\varphi$ of the angles $\theta$ and $\varphi$ (Fig.5.4). The element being considered is a rectangle with the sides $r\mathrm{d}\theta$ and $r\sin\theta\mathrm{d}\varphi$. Thus

$$\mathrm{d}S = r^2\sin\theta\mathrm{d}\theta\mathrm{d}\varphi \tag{5.14}$$

The expression obtained gives an element of the surface $r$ = const in a spherical system of coordinates.

Dividing Eq. (5.14) by $r^2$ we shall find the element of the solid angle corresponding to the angle intervals from $\theta$ to $\theta + \mathrm{d}\theta$ and from $\varphi$ to $\varphi + \mathrm{d}\varphi$.

$$\mathrm{d}\Omega_{\theta}, \varphi = \sin\theta\mathrm{d}\theta\mathrm{d}\varphi \tag{5.15}$$

Two spheres of radius $r$ and $r + \mathrm{d}r$, two cones with the apex angles $\theta$ and $\theta + \mathrm{d}\theta$, and two planes forming the angles $\varphi$ and $\mathrm{d}\varphi$ with $P_0$ separate in space a rectangular parallelepiped with the sides $r\mathrm{d}\theta$, $r\sin\theta\mathrm{d}\varphi$ and $\mathrm{d}r$ (see Fig. 5.4). The volume of this parallelepiped

$$\mathrm{d}V = r^2\sin\theta\mathrm{d}r\mathrm{d}\theta\mathrm{d}\varphi \tag{5.16}$$

is an element of volume in a spherical system of coordinates (the volume corresponding to an increase in the coordinates $r$, $\theta$ and $\varphi$ by $\mathrm{d}r$, $\mathrm{d}\theta$, and $\mathrm{d}\varphi$).

Passing over from deltas to differentials in Eq. (5.13) and introducing Eq. (5.15) for $\mathrm{d}\Omega$, we arrive at the expression

$$\mathrm{d}N_{\theta,\varphi} = N\frac{\mathrm{d}\Omega_{\theta,\varphi}}{4\pi} = N\frac{\sin\theta\mathrm{d}\theta\mathrm{d}\varphi}{4\pi} \tag{5.17}$$

The subscripts $\theta$ and $\varphi$ of $\mathrm{d}N$ indicate that we have in view the molecules whose directions of motion correspond to the angle intervals from $\theta$ to $\theta + \mathrm{d}\theta$ and from $\varphi$ to $\varphi + \mathrm{d}\varphi$.

## 5.3 NUMBER OF COLLISIONS OF MOLECULES WITH A WALL

Let us consider a gas in equilibrium confined in a vessel. We shall take an element $\Delta S$ of the surface of the vessel and count the number of collisions of molecules with this element $\Delta S$ during the time $\Delta t$.

Let us separate from the $N$ molecules in the vessel those $\mathrm{d}N_v$ molecules whose velocities have magnitudes ranging from $v$ to $v + \mathrm{d}v$.

Of these molecules, the number of molecules whose directions of motion are confined within the solid angle $\mathrm{d}\Omega$, equals

$$\mathrm{d}N_{v,\theta,\varphi} = \mathrm{d}N_v\frac{\mathrm{d}\Omega_{\theta,\varphi}}{4\pi} \tag{5.18}$$

[see Eq. (5.17)]. Of the molecules separated in this way, the ones confined in an oblique cylinder with the base $\Delta S$ and the altitude $(v\cos\theta)\Delta t$ (Fig. 5.5) will fly during the time $\Delta t$ up to the area $\Delta S$ and collide with it. The number of these molecules is

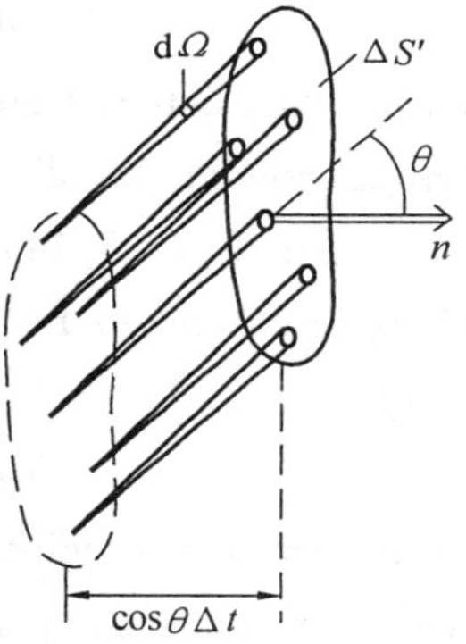


Fig.5.5

$$\mathrm{d}\nu_{v,\theta,\varphi} = \mathrm{d}N_v \frac{\mathrm{d}\Omega_{\theta,\varphi}}{4\pi} \frac{\Delta S(v \cos \theta)\Delta t}{V} \tag{5.19}$$

( $V$ is the volume of the vessel).

To obtain the total number of collisions of the molecules with the area $\Delta S$, we must summate Eq. (5.19) over the solid angle $2\pi$ (corresponding to changes in $\theta$ from 0 to $\pi/2$ and changes in $\varphi$ from 0 to $2\pi$) and over the velocities ranging from 0 to $v_{\max}$ where $v_{\max}$ is the maximum velocity the molecules can have in the given conditions (see the preceding section).

We shall begin with summation over the directions. For this purpose, we shall write $\mathrm{d}\Omega$ in the form $\theta \mathrm{d}\theta \mathrm{d}\varphi$ [see Eq. (5.15)] and integrate Eq. (5.19) with respect to $\theta$ within the limits from 0 to $\pi/2$ and with respect to $\varphi$ within the limits from 0 to $2\pi$:

$$\mathrm{d}\nu_v = \frac{\mathrm{d}N_v v \Delta S \Delta t}{4\pi V} \int_0^{\pi/2} \sin \theta \cos \theta \mathrm{d}\theta \int_0^{2\pi} \mathrm{d}\varphi$$

Integration with respect to $\mathrm{d}\varphi$ gives $2\pi$, and the integral with respect to $\mathrm{d}\theta$ equals 1/2. Hence,

$$\mathrm{d}\nu_v = \frac{\mathrm{d}N_v v \Delta S \Delta t}{4V} \tag{5.20}$$

This expression gives the number of times the molecules flying in the directions confined within the solid angle $2\pi$ and having velocities from $v$ to $v + \mathrm{d}v$ collide with the area $\Delta S$ during the time $\Delta t$.

Summation over the velocities gives the total number of collisions of the molecules with the area $\Delta S$ during the time $\Delta t$:

$$\nu_{\Delta S, \Delta t} = \frac{\Delta S \Delta t}{4V} \int_0^{v_{\max}} v \mathrm{d}N_v \tag{5.21}$$

The expression

$$\frac{1}{N} \int_0^{v_{\max}} v \mathrm{d}N_v$$

is the mean value of the speed $v$. Substituting the product $N\langle v \rangle$ for the integral in Eq. (5.21), we find that

$$\nu_{\Delta S,\Delta t} = \frac{\Delta S \Delta t}{4V} N\langle v\rangle = \frac{1}{4}\Delta S \Delta t n\langle v\rangle \qquad (5.22)$$

Here $n = N/V$ is the number of molecules of a gas in unit volume.

Finally, dividing Eq. (5.22) by $\Delta S$ and $\Delta t$, we shall find the number of collisions of the gas molecules with a unit surface area of the wall in unit time:

$$\nu = \frac{1}{4} n\langle v\rangle \qquad (5.23)$$

The result obtained signifies that the number of collisions is proportional to the number of molecules per unit volume (the "concentration" of the molecules) and to the mean value of the speed of the molecules (and not their velocity – the mean value of the velocity vector of the molecules for equilibrium of a gas is zero). We must note that the quantity $\nu$ in Eq. (5.23) is the density of the stream of molecules striking the wall.

Let us consider an imaginary unit area in a gas. If the gas is in equilibrium, the same number of molecules will fly through this area in both directions on an average. The number of molecules flying in each direction in unit time is also determined by Eq. (5.23).

Equation (5.23) can be obtained with an accuracy up to the numerical coefficient with the aid of the following simplified reasoning. Let us assume that the gas molecules travel only in three mutually perpendicular directions. If our vessel contains $N$ molecules, then at any moment $N/3$ molecules will travel in each of these directions. One half of them (i.e. $N/6$ molecules) will travel in a given direction to one side, and the other half to the other side. Hence, 1/6 of the molecules travel in the direction we are interested in (for example, along a normal to the given element $\Delta S$ of the vessel's wall).

Let us also assume that all the molecules travel with the same speed equal to $\langle v\rangle$. Therefore, during the time $\Delta t$, the wall element $\Delta S$ will be readied by all the molecules moving toward it that are inside a cylinder

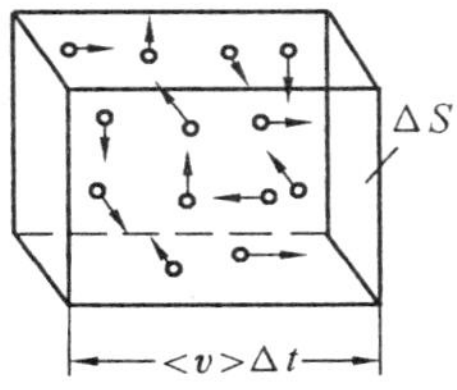


Fig.5.6

with the base $\Delta S$ and the altitude $\langle v\rangle\Delta t$ (Fig. 5.6). The number of these molecules is $\Delta\nu = (n/6)\Delta S\langle v\rangle\Delta t$. Accordingly, the number of collisions with a unit area in unit time will be

$$\nu = \frac{1}{6}n\langle v\rangle \qquad (5.24)$$

The expression obtained differs from Eq. (5.23) only in the value of the numerical factor (1/6 instead of 1/4).

Retaining our assumption on the motion of the molecules in three mutually perpendicular directions, but negating the assumption on the molecules having identical speeds, we must separate from among the molecules in unit volume those $dn_v$ molecules whose speed range from $v$ to $v + dv$. The number of molecules having such speeds and reaching the area $\Delta S$ during the time $\Delta t$ is

$$d\nu_v = \frac{1}{6}dn_v\Delta Sv\Delta t \qquad (5.25)$$

We get the total number of collisions by integrating Eq. (5.25) with respect to speeds:

$$\Delta\nu = \int d\nu_v = \frac{1}{6}\Delta S\Delta t\int_0^{v_{max}} v\,dn_v = \frac{1}{6}\Delta S\Delta tn\langle v\rangle$$

Finally, dividing $\Delta\nu$ by $\Delta S$ and $\Delta t$, we get Eq. (5.24). Thus, our assumption that the molecules have identical speeds does not affect the result obtained for the number of collisions of the molecules with the wall. As we shall see in the following section, however, this assumption changes the result of pressure calculations.

## 5.4 PRESSURE OF A GAS ON A WALL

The walls of a vessel containing a gas are continuously bombarded by its molecules. The result is that the wall element $\Delta S$ receives a momentum during one second that equals the force acting on this element. The ratio of this force to the area $\Delta S$ gives the pressure exerted by the gas on the walls of the vessel. The pressure of the gas on different portions of the vessel walls is the same owing to the chaotic nature of motion of the molecules (naturally, provided that the gas is in an equilibrium state).

If we assume that the molecules rebound from a wall according to the law of mirror reflection ($\theta_{refl} = \theta_{inc}$) and the magnitude of the velocity of a molecule does not change, then the momentum imparted by a molecule to the wall upon colliding with it will be $2mv\cos\theta$ (Fig. 5.7), where $m$ is the mass of a molecule. This momentum is directed along a normal [see Eq. (5.19)] imparts a momentum of $2mv\cos\theta$ the wall, to the area. Each of the $d\nu_{v,\theta,\varphi}$ molecules and all these molecules impart a momentum of

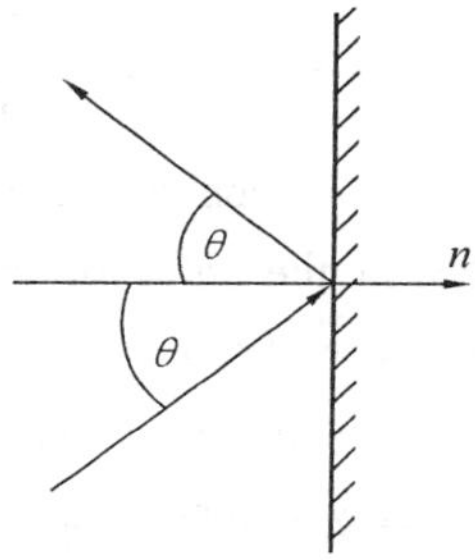


**Fig. 5.7**

$$dK_{v,\theta,\varphi} = 2mv\cos\theta d\nu_{v,\theta,\varphi} = dN_v \frac{d\Omega_{\theta,\varphi}}{4\pi} \frac{2mv^2\cos^2\theta\Delta S\Delta t}{V}$$

(We have used the symbol $K$ for the momentum in stead of the previously used symbol $p$ to avoid confusion—the latter symbol stands for pressure here.)

Summation of the expression obtained by directions within the limits of the solid angle $2\pi$ (corresponding to changes in $\theta$ from 0 to $\pi/2$ and changes in $\varphi$ from 0 to $2\pi$) gives the momentum imparted by the molecules whose velocities have magnitudes ranging from $v$ to $v + dv$:

$$dK_v = dN_v \frac{2mv^2\Delta S\Delta t}{4\pi V}\int_0^{\pi/2}\cos^2\theta\sin\theta d\theta\int_0^{2\pi}d\varphi$$

[we have introduced Eq. (5.15) for $d\Omega$]. Integration with respect to $d\varphi$ yields $2\pi$, and the integral with respect to $d\theta$ is 1/3. Hence,

$$dK_v = dN_v \frac{mv^2\Delta S\Delta t}{3V}$$

Integrating this expression with respect to velocities from 0 to $v_{max}$, we get the total momentum imparted to the area $\Delta S$ during the time $\Delta t$:

$$\Delta K = \frac{m\Delta S\Delta t}{3V}\int_0^{v_{max}} v^2 \mathrm{d}N_v \tag{5.26}$$

The expression

$$\frac{1}{N}\int_0^{v_{max}} v^2 \mathrm{d}N_v$$

is the mean value of the square of the velocity of the molecules. Substituting the product $N\langle v^2\rangle$ for the integral in Eq. (5.26), we find that

$$\Delta K = \frac{m\Delta S\Delta t}{3V}N\langle v^2\rangle = \frac{1}{3} nm\langle v^2\rangle\Delta S\Delta t$$

($n = N/V$ is the number of molecules in unit volume). Finally, dividing this expression by $\Delta S$ and $\Delta t$, we obtain the pressure of a gas on the walls of the vessel containing it:

$$p = \frac{1}{3}nm\langle v^2\rangle = \frac{2}{3}n\frac{m\langle v^2\rangle}{2} \tag{5.27}$$

We have assumed that all the molecules have the same mass. We can therefore put it inside the sign of the mean quantity. As a result, Eq. (5.27) acquires the form

$$p = \frac{2}{3}n\langle\frac{mv^2}{2}\rangle = \frac{2}{3}n\langle\varepsilon_{tr}\rangle \tag{5.28}$$

where $\langle\varepsilon_{tr}\rangle$ is the mean value of the kinetic energy of translation of the molecules.

Let us obtain an expression for the pressure proceeding from the simplified notions that led us to Eq. (5.24). According to these notions, each molecule imparts a momentum of $2m\langle v\rangle$ to the wall it collides with. Multiplying this momentum by the number of collisions [see Eq. (5.24)], we get the momentum imparted to a unit area in unit time, i.e. the pressure. We thus obtain the equation

$$p = \frac{1}{6}n\langle v\rangle 2m\langle v\rangle = \frac{1}{3}nm\langle v\rangle^2 \tag{5.29}$$

This equation differs from Eq. (5.27) in that it contains the square of the mean velocity $\langle v\rangle^2$ instead of the mean square of the velocity $\langle v^2\rangle$. We shall see on a later page that these two quantities differ from each other, i.e. $\langle v^2 \rangle \neq \langle v \rangle^2$.

In a more accurate calculation, we must multiply the number of molecules determined according to Eq. (5.25) by $2mv$ and then summate over all the $v$'s. As a result, we get the momentum imparted to the area $\Delta S$ during the time $\Delta t$:

$$\Delta K = \int_0^{v_{\max}} \frac{1}{6} \mathrm{d}n_v \Delta S \Delta t \cdot 2mv = \frac{1}{3} \Delta S \Delta t m \int_0^{v_{\max}} v^2 \mathrm{d}n_v = \frac{1}{3} \Delta S \Delta t n m \langle v^2 \rangle$$

Dividing this equation by $\Delta S$ and $\Delta t$ we get Eq. (5.27) for the pressure. Thus, on the basis of our simplified notion of the molecules traveling in three mutually perpendicular directions, we have obtained an exact expression for the pressure. The explanation is that this simplification leads on the one hand to diminishing of the number of collisions of the molecules with the wall $\frac{4}{6} n \langle v \rangle$ instead of $\frac{1}{4} n \langle v \rangle$, [see Eqs. (5.23) and (5.24)], and on the other to overstating of the momentum transmitted to the wall in each collision. In our simplified derivation, we assumed that the wall receives a momentum of $2mv$ upon each collision. Actually, however, the magnitude of the momentum imparted to the wall depends on the angle $\theta$ and as a result the mean momentum imparted in one collision is $\frac{4}{3} mv$. In the long run, both inaccuracies mutually compenate each other and, notwithstanding the simplified nature of our derivation, we obtain an exact expression for the pressure.

## The Further Study:

### Mean Energy of Molecules

Let us write Eq. (5.28) for the pressure obtained in the preceding section and the equation of state of an ideal gas next to each other:

$$p = \frac{2}{3} n \langle \varepsilon_{\mathrm{tr}} \rangle; \quad p = nkT$$

A comparison of these equations shows that

$$\langle \varepsilon_{tr} \rangle = \frac{3}{2} kT \tag{5.30}$$

We have thus arrived at an important conclusion: the absolute temperature is a quantity proportional to the mean energy of translation of molecules. Only gas molecules have translation. For liquids and solids, the mean energy of the molecules is proportional to the absolute temperature only when the motion of the molecules can be treated classically. In the quantum region, the mean energy of the molecules stops being proportional to the absolute temperature.

Equation (5.30) is remarkable in that the mean energy is found to depend only on the temperature and is independent of the mass of a molecule.

Since $\langle \varepsilon_{tr} \rangle = \langle mv^2/2 \rangle = (m/2)\langle v^2 \rangle$ it follows from Eq. (5.30) that

$$\langle v^2 \rangle = \frac{3kT}{m} \tag{5.31}$$

Representing $v^2$ in the form of the sum of the squares of the velocity components, we can write:

$$\langle v^2 \rangle = \langle v_x^2 \rangle + \langle v_y^2 \rangle + \langle v_z^2 \rangle$$

Owing to all the directions of motion having equal rights, the following equation is obeyed:

$$\langle v_x^2 \rangle = \langle v_y^2 \rangle = \langle v_z^2 \rangle$$

Taking this into account, we find that

$$\langle v_x^2 \rangle = \frac{1}{3}\langle v^2 \rangle = \frac{kT}{m} \tag{5.32}$$

Equation (5.30) determines the energy of only the translation of a molecule. In addition to translation, however, rotation of a molecule and vibrations of the atoms in the molecule are possible. Both these kinds of motion are associated with a certain store of energy. The latter can be determined by the theorem on the equal distribution of the energy by the degrees of freedom of a molecule established by statistical physics.

The number of degrees of freedom, of a mechanical system is defined as the number of independent quantities by means of which we can set the position of the system. Thus, the position of a point particle in space is determined completely by setting the values of three of its coordinates (for example, the Cartesian coordinates $x$, $y$, $z$, or the spherical coordinates $r$, $\theta$, $\varphi$, etc.). Accordingly, a point particle has three degrees of freedom.

The position of a perfectly rigid body can he determined by setting three coordinates of its center of mass ($x$, $y$, $z$), the two angles $\theta$ and $\varphi$ indicating the direction of an axis associated with the body and passing through its center of mass (Fig. 5.8), and, finally, the angle $\psi$ determining the direction of a second axis associated with the body and perpendicular to the first one. Hence, a perfectly rigid body has six degree freedom. A change in the coordinates of the center of mass with the angles $\theta$, $\varphi$ and $\psi$ remaining constant is due to translation of a rigid body. Therefore, the relevant degrees of freedom are called translational. A change in any of the angles $\theta$, $\varphi$, $\psi$ with an unchanging position of the center of mass is due to rotation of a body, and in this connection the corresponding degrees of freedom are called rotational. Hence, of the six degrees of freedom of a perfectly rigid body, there are translational and three rotational.

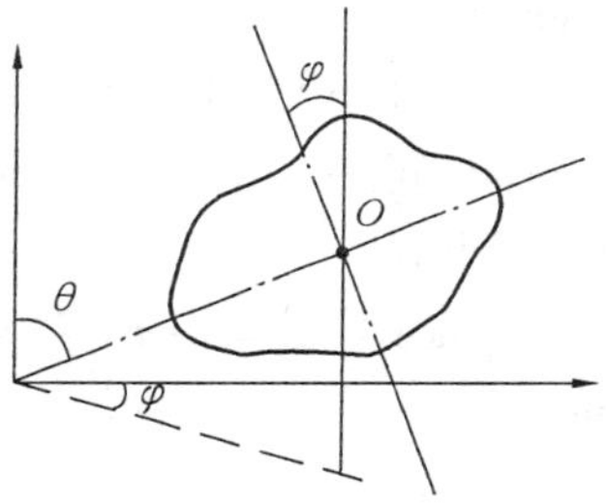

**Fig.5.8**

A system of $N$ point particles between which there are no rigid constraints has $3N$ degrees of freedom (the position of each of the $N$ particles must be set by three coordinates). Any rigid constraint establishing an unchanging mutual arrangement of two particles reduces the number of degrees of freedom by one. For example, if a system consists of two point particles with a constant distance $l$ between them (Fig.5.9), then the number of degrees of freedom of the system is live. Indeed, in this case, the following relation holds between the coordinates of the particles:

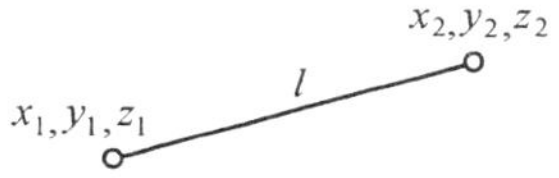


**Fig.5.9**

$$(x_1 - x_2)^2 + (y_1 - y_2)^2 + (z_1 - z_2)^2 = l^2 \qquad (5.33)$$

owing to which the coordinates will not be independent: it is sufficient to set any five coordinates, and the sixth one will be determined by condition (2.33). To classify these five degrees of freedom, we shall note that the position of a system formed by two rigidly connected point particles can be determined as follows: we can set, the three coordinates of the center of mass of the system (Fig .5.10) and the two angles $\theta$ and $\varphi$ that determine the direction in space of the axis of the system (i.e. the straight line passing through both points). It thus follows that three degrees of freedom will be translational and two rotational. The latter correspond to rotation about two mutually perpendicular axes $O'O'$ and $O''O''$ that are at right angles to axis $OO$ of the system (Fig.5.11). Rotation about axis $OO$ is deprived of meaning for point particles.

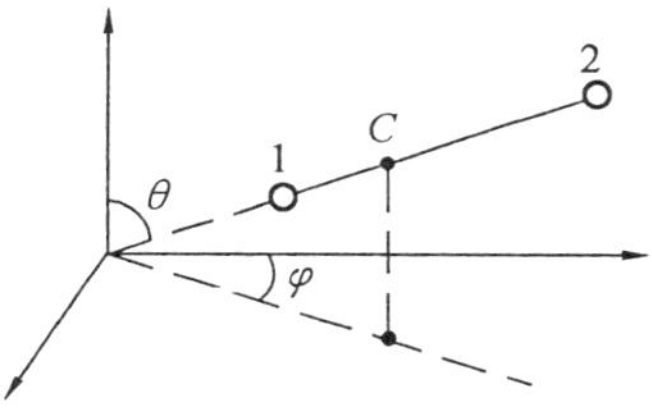


**Fig.5.10**

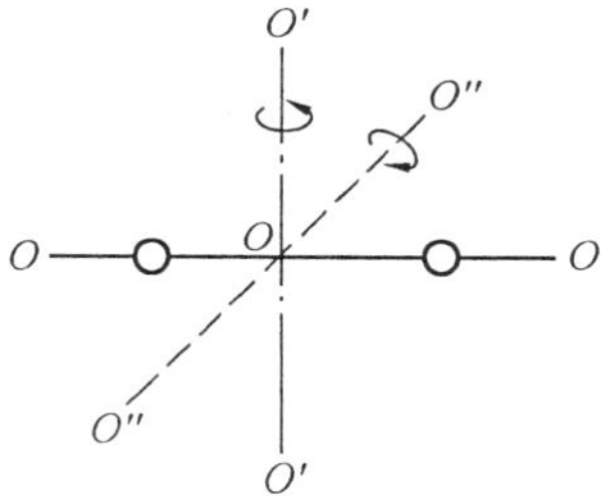


**Fig.5.11**

If two point particles are connected by an elastic constraint instead of a rigid one (i.e. so that any change in the equilibrium distance $r_o$ between the particles results in the setting up of forces tending to establish the initial distance between the particles), then the number of degrees of freedom will be six. The position of the system in this case can be determined by setting the three coordinates of the center of mass (Fig.5.12), the two angles $\theta$ and $\varphi$, and the distance $r$ between the particles. Changes in $r$ correspond to vibrations in the system, consequently this degree of freedom is called vibrational. Thus, the system considered has three translational, two rotational, and one vibrational degree of freedom.

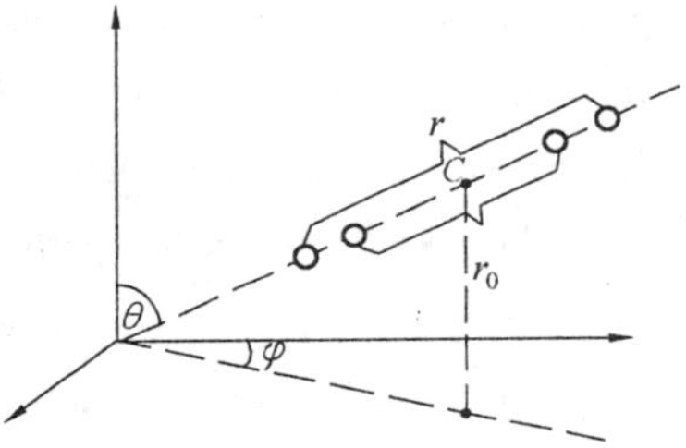

**Fig.5.12**

Let us consider a system consisting of $N$ point particles elastically connected one another. Such a system has $3N$ degrees of freedom. The particles have an equilibrium configuration corresponding to a minimum potential energy of the system. The equilibrium configuration is characterized by quite definite mutual distances between the particles. If the particles are brought out of their positions corresponding to equilibrium configuration, vibrations appear in the system. The position of the system can be determined by setting the position of the equilibrium configuration and the quantities characterizing the displacements of the particles from their equilibrium positions. The latter quantities correspond to the vibrational degrees of freedom.

The position of equilibrium configuration, like that of a perfectly rigid body, is determined by six quantities which three translational and three rotational degrees of freedom correspond to. The number of vibrational degrees of freedom is thus $3N-6$.

Experiments on measuring the heat capacity of gases have shown that atoms must be treated as point particles in determining the number of degrees of freedom of a molecule. Consequently, three translational degrees of freedom should be ascribed to a monatomic molecule. The degrees of freedom ascribed to a diatomic molecule depend on the nature of the bond between the atoms. They include either three translational and two rotational degrees of freedom (with a rigid bond), or, apart from these five, another vibrational degree of freedom (with an elastic bond). A triatomic molecule with rigid bonds has three translational and three rotational degrees of freedom, etc.

We must note that no matter how many degrees of freedom a molecule has, three of them are translational. Since none of the translational degrees of freedom of a molecule has priority over the other two, an identical energy should fall to each of them on an average. This energy is one-third of the value given by Eq. (5.30), i.e. $\frac{1}{2}kT$.

The **equipartition principle** is derived in classical statistical physics. It states that an identical kinetic energy equal to $\frac{1}{2}kT$ resides on the average in any degree of freedom.

According to this principle, the mean energy of one molecule $\langle \varepsilon \rangle$ will be the greater (at the same temperature), the more complex is the molecule and the more degrees of freedom it has. In determining $\langle \varepsilon \rangle$, we must take into account that a vibrational degree of freedom must have an energy capacity that is twice the value for a translational or rotational one. The explanation is that translation and rotation of a molecule are associated with the presence of only kinetic energy, whereas vibration is associated with the presence of both kinetic potential energy; for a harmonic oscillator, the mean $S$ of the kinetic and the potential energy is the same. Hence, two halves of $kT$ must reside in each vibrational degree of freedom – one in the form of kinetic energy and one in the form of potential energy.

Thus, the mean energy of a molecule should be

$$\langle \varepsilon \rangle = \frac{i}{2} kT \tag{5.34}$$

where $i$ is the sum of the number of translational, the number of rotational, and the double number of vibrational degrees of freedom of a molecule:

$$i = n_{\text{tr}} + n_{\text{rot}} + 2n_{\text{vib}} \tag{5.35}$$

For molecules with a rigid bond between their atoms, $i$ coincides with the number of degrees of freedom of a molecule.

Molecules an ideal gas do not interact with one another. We can ore find the internal energy of one mole of an ideal gas by multiplying the Avogadro constant by the mean energy of one molecule:

$$U_m = N_{\text{A}} \langle \varepsilon \rangle = \frac{i}{2} N_{\text{A}} kT = \frac{i}{2} RT \tag{5.36}$$

A comparison of this equation with Eq. (4.28) shows that

$$C_V = \frac{i}{2} R \tag{5.37}$$

With a view to Eq. (4.33), we find that

$$C_p = \frac{i+2}{2} R \tag{5.38}$$

Hence,

$$\gamma = \frac{C_p}{C_V} = \frac{i+2}{i} \tag{5.39}$$

Thus, the quantity $\gamma$ is determined by the number and the nature of degrees of freedom of a molecule.

Table (5.1) gives the values of $C_V$, $C_p$ and $\gamma$ obtained for different speeds of molecules by Eqs. (5.37), (5.38), and (5.39). Table(5.2) compares the theoretical results with experimental data. The theoretical values have been obtained (except for the case indicated in the footnote to the table) on the assumption that the molecules are rigid; the experimental values have been obtained for temperatures close to room temperature.

**Table 5.1**

| Molecule | Nature of interatomic bond | Number of degrees of freedom | | | $i$ | $C_v$ | $C_p$ | $\gamma_i$ |
|---|---|---|---|---|---|---|---|---|
| | | translational | rotational | vibrational | | | | |
| Monatomic | — | 3 | — | — | 3 | $\frac{3}{2}R$ | $\frac{5}{2}R$ | 1.67 |
| Diatomic | Rigid | 3 | 2 | — | 5 | $\frac{5}{2}R$ | $\frac{7}{2}R$ | 1.40 |
| Ditto | Elastic | 3 | 2 | 1 | 7 | $\frac{7}{2}R$ | $\frac{9}{2}R$ | 1.29 |
| With three or more atoms | Rigid | 3 | 2 | — | 6 | $\frac{6}{2}R$ | $\frac{8}{2}R$ | 1.33 |

It should seem to follow from Table 5.2 that agreement between theory and experiments is quite satisfactory, at any rate for mon – and diatomic molecules. Actually, however, matters are different. According to the theory we have considered, the heat capacities of gases

**Table 5.2**

| Gas | Number of atoms in molecule | $C_V \times 10^{-3}$ | | $C_p \times 10^{-3}$ | | $\gamma$ | |
|---|---|---|---|---|---|---|---|
| | | theor. | exper. | theor. | exper. | theor. | exper. |
| Helium(He) | 1 | 12.5 | 12.5 | 20.8 | 20.9 | 1.67 | 1.67 |
| Oxygen($O_2$) | 2 | 20.8 | 20.9 | 29.1 | 29.3 | 1.40 | 1.40 |
| Carbon monoxide(CO) | 2 | 20.8 | 21.0 | 29.1 | 29.3 | 1.40 | 1.40 |
| Water vapo (ur$H_2O$) | 3 | 25.0<br>33.2 | 27.8 | 33.2<br>41.5 | 36.2 | 1.33<br>1.25 | 1.31 |

ought to be integral multiples of $R/2$ because the number of degrees of freedom can only be integral. Therefore, even small deviations of $C_V$ and $C_p$ from values that are multiples of $R/2$ have fundamental significance.

Examination of the table shows that such deviations, exceeding the possible errors of measurements, are encountered.

The discrepancies between theory and experiments become especially striking if we turn to the temperature dependence of the heat capacity. Figure 5.13 shows a curve of the temperature dependence of the molar heat capacity $C_V$ obtained experimentally for hydrogen. The heat capacity should be independent of the temperature according to theory. A glance at the figure shows that this holds only within the limits of separate temperature intervals, and that within different intervals the heat capacity has values corresponding to different numbers of degrees of freedom of a molecule. Thus, on portion 1 – 1′ we have $C_V = \frac{3}{2}R$ . This signifies that a molecule behaves like a system having only translational degrees of freedom. On portion 2 – 2′, we have $C_V = \frac{5}{2}R$. Hence, at temperatures corresponding to this portion of the curve, in addition to the three translational degrees of freedom manifesting themselves at lower temperatures, two rotational ones appear in a molecule. Finally, at sufficiently high temperatures, $C_V$ becomes equal to $\frac{7}{2}R$, which points to the presence of vibrations of a molecule at these temperatures. Between these intervals, the heat capacity monotonously grows with increasing temperature, i.e. corresponds, as it were, to a fractional varying number of degrees of freedom.

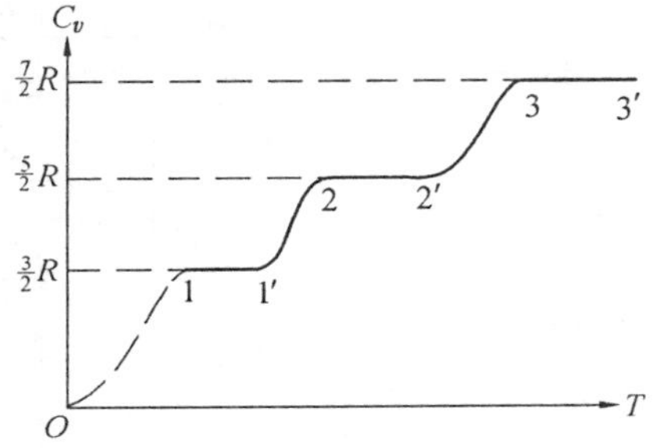


**Fig.5.13**

Thus, the number of degrees of freedom of a molecule manifesting itself in the heat capacity depends on the temperature. At low temperatures, only translation of the molecules is observed. At higher temperatures, rotation of the molecules is observed in addition to translation. And, finally, at still higher temperatures, vibrations of the molecules are also added to the first two kinds of motion. As indicated by the monotonous nature of the heat capacity curve here not all the molecules at a time are involved in rotation, and then in vibration. First rotation, for example, begins to be observed only in a small traction of the molecules. This fraction grows with elevation of the temperature, and in the long run when a definite temperature is reached, virtually all the molecules will be involved in rotation. Matters are similar for vibration of the molecules.

This behavior of the heat capacity is explained by quantum mechanics. The quantum theory has established that the energy of rotation and vibration of molecules is quantized. This signifies that the energy of rotation and that of vibration of a molecule cannot have any values, but only discrete ones (i.e. values differing from one another by a finite amount). Consequently the energy associated with these kinds of motion can change only in jumps. Such restrictions do not exist for the energy of translation.

The intervals between separate allowed values of the energy (or, in accordance with the adopted terminology, between energy levels) are about an order greater for vibration than for rotation. A simplified diagram of the rotational and vibrational levels of a diatomic molecule is given in Fig. 5.14. (The distances between the rotational levels are actually not the same, but this is of no significance for the question being considered.)

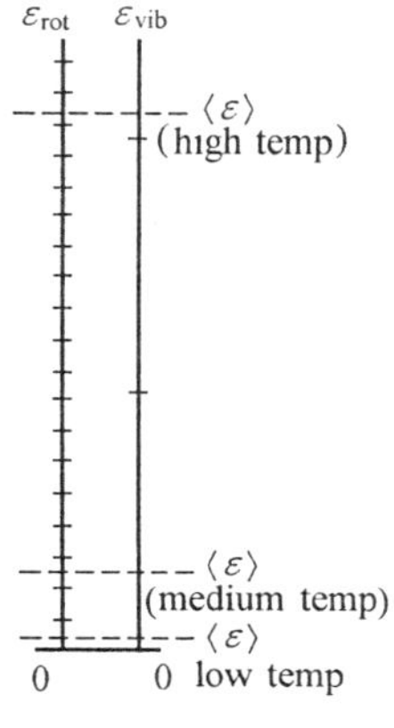


**Fig. 5.14**

We noted in Sec. 5.2 that the velocities of molecules are mainly grouped near a most probable value. Accordingly, the predominating part of molecules have energies close to the mean value $\langle \varepsilon \rangle$, and only a small part of the molecules have energies considerably exceeding $\langle \varepsilon \rangle$. Hence, for an appreciable part of the molecules to be involved in rotation or vibration, their mean energy must be sufficiently high in comparison with the distance between the allowed levels of the relevant energy.

Let us take such a low temperature that the mean energy of a molecule $\langle \varepsilon \rangle$ is considerably lower than the first allowed value of the rotational energy (see the bottom dash line in Fig.5.14). Now only an insignificant part of all the molecules will be involved in rotation, so that the molecules of the gas will virtual have only translation. Small changes in the temperature will result in changes only in the energy of translation, and the heat capacity of the gas will accordingly by $\frac{3}{2}R$ (see $1 - 1'$ on the curve, depicted in Fig.5.13).

Elevation of the temperature is attended by ah increase in $\langle \varepsilon \rangle$ so that a constantly growing part of the molecules will be involved in rotation. Portion $1' - 2$ of the curve $m$ Fig.5.13 corresponds to this process.

After all the molecules begin to participate in rotation, the horizontal portion $2 - 2'$ commences. At temperatures corresponding to it, the value of $\langle \varepsilon \rangle$ is still considerably lower than the distance between the allowed levels of vibrational energy. As a result, vibration of the molecules will virtually be absent. With a further elevation of the temperature, the molecules will begin to vibrate in greater and greater numbers, which transition portion $2' - 3$ on the heat capacity curve corresponds to. Finally, at a sufficiently high temperature, all the molecules will be involved in vibration, and the heat capacity will become equal to $\frac{7}{2}R$.

The classical theory of heat capacity is thus approximately correct only for separate temperature intervals. A different number of degrees of freedom of a molecule corresponds to each interval.

### Summary of Key Terms

**First law of thermodynamics** A restatement of the law of energy conservation, usually as it applies to systems involving changes in temperature. The heat added to a system equals an increase in internal energy plus external work done by the system.

**Adiabatic process** A process, usually of expansion or compression, wherein no heat enters or leaves a system.

**Second law of thermodynamics** Heat will never spontaneously flow from a cold object to a hot object. Also, no machine can be completely efficient in converting energy to work; some input energy is dissipated as heat. And finally, all systems tend to become more and more disordered as time goes by.

**Heat engine** A device that changes internal energy to mechanical work.

**Entropy** A measure of the disorder of a system. Whenever energy freely transforms from one form to another, the direction of transformation is toward a state of greater disorder and therefore toward one of greater entropy.

## Reading Materials:

### The Greenhouse Effect

If the earth continually receives energy from the sun, you may wonder why the Earth doesn't get hotter or why its average temperature does not increase. To maintain a relatively constant long-term average temperature, the Earth must lose energy, which on the average is equal to the amount it receives. This is accomplished through the re-radiation of energy back into space. Atmospheric effects on how this is done are important in preventing large variations in daily temperature. On the moon, which has no atmosphere, daily temperature variations range from about 100℃ (212°F, the boiling point of water) on the day or Sun side to −173℃ (−280°F) on the dark side.

Incoming solar radiation warms the atmosphere and the surface of the Earth, and the warm Earth re-radiates energy in the form of infrared radiation. The gases of the atmosphere water vapor and carbon dioxide ($CO_2$), are "selective absorbers". That is, they allow the visible incoming sunlight to pass through, but they absorb or trap certain infrared radiations. This atmospheric absorption helps to retain the Earth's energy, so we don't have the daily temperature fluctuations as on the moon. Clouds (water droplets) also assist in maintaining the Earth's warmth by absorbing infrared radiation. In the absence of cloud coverage, nights are "cold and clear".

Hence, the atmospheric faces have a "thermostatic" effect in maintaining the Earth's daily temperature variations. We call this process the greenhouse effect. Glass has absorption properties similar to those of the atmospheric gases. As used in a greenhouse, the glass allows the visible sunlight to pass through, then blocks or absorbs the infrared radiation. actually, in this case the warmth is primarily due to the prevention of the escape of warm air heated by the ground within the glass enclosure. The temperature of a greenhouse in the summer is controlled by painting the glass panels white, which reflects the sunlight, and opening panels to allow the hot air to escape.

The interior of a closed greenhouse is quite warm, even on a cold day. You have probably experienced the "greenhouse" effect in an automobile on a cold, sunny day.

## Scientist: Johannes Kepler

Tycho Brahe was fortunate. In 1576 the king of Denmark financed a fine observatory for him on the island of Hven northeast of Copenhagen.

There for the next 20 years Brahe made extensive observations of the stars and planets. These were done without the benefit of a telescope, which had not yet been invented. Even so, Brahe's unaided – eye observations were more accurate than previous ones because of better instrumentation. As a result, he is considered to be one of the greatest practical astronomers of modern times. On the basis of observations that supported a mistaken ideas of the time, Brahe tried to convert Galileo, a contemporary, from his belief in the Copernican heliocentric theory, but to no avail.

Anew king withdrew Brahe's royal support, and Brahe left Denmark, taking his instruments and records with him. In 1599 he arrived in Prague, where the emperor provided him support in the capacity of the court mathematician. Brahe invited and was joined by a young German mathematician and astronomer, Johannes Kepler, in 1600.

Kepler, who had just celebrated his 29th birthday, had been a professor of mathematics at the University of Tubingen, where he had become acquainted with the heliocentric concept of Copernicus. His interest in planetary motion was in part associated with astrological work and predictions, for which he was responsible at the university. The positions of stars and planets supposedly exerted an influence on such predictions.

Because of religious unrest between Catholics and Protestants, Kepler and his wife had to leave the university in 1600 m despite his wife's wealthy Protestant influences. Friends suggested that Tycho Brahe might help and he did. (Because of his publications Kepler had corresponded with Brahe and Galileo). And so, Kepler went to assist Brahe in Prague.

Brahe died the next year, and Kepler was appointed his successor as court mathematician. Kepler inherited Brahe's records of the positions of the planets that he had made over many years. These were to prove a storehouse of information and to provide the basis for Kepler's laws of planetary motion for which he is famous.

Applied initially only to Mars, the first two of Kepler's three laws were announced in 1609. (In the same year, Galileo built his first telescope.) This was done in a book in which he also theorized that some kind of force on the planets emanated from the Sun. He presumed the orbital motions of the planets to be in some way associated with the rotation of the Sun. Although incorrect, it was one of the first attempts to associate force with the motions of the planets.

Kepler's attention then turned to the use of the newly invented telescope. Having had previous interest in optics, he published a work in 1611 on telescope design that later came into wide use. With intervening theological writings and a treatise on comets, in 1619, ten years after the publication of his first two laws, Kepler published his third law of planetary motion in the work De Harmonica Mundi. In this work, Kepler proposed a mathematical concept of "harmony" in the solar system.

His work on documenting the position of the planets was to continue for another ten years. Kepler died in 1630, having made major contributions to astronomy, among them the dethroning of the Earth as the center of the universe and solar system.

PART 3

OPTICS

# 6

# Interference of Light

## 6.1 INTERFERENCE OF LIGHT WAVES

Let us assume that two waves of the same frequency, being superposed on each other, produce oscillations of the same direction, namely,

$$A_1\cos(\omega t + \alpha_1); \quad A_2\cos(\omega t + \alpha_2);$$

at a certain point in space. The amplitude of the resultant oscillation at the given point is determined by the expression

$$A^2 = A_1^2 + A_2^2 + 2A_1A_2\cos\delta$$

where $\delta = \alpha_2 - \alpha_1$

If the phase difference $\delta$ of the oscillations set up by the waves remains constant in time, then the waves are called coherent.

The phase difference $\delta$ for incoherent waves varies continuously and takes on any values with an equal probability. Hence, the time averaged value of cos $\delta$ equals zero. Therefore

$$< A^2 > = < A_1^2 > + < A_2^2 >$$

We thus conclude that the intensity observed upon the superposition of incoherent waves equals the sum of the intensities produced by each of the waves individually:

$$I = I_1 + I_2 \tag{6.1}$$

For coherent waves, cos $\delta$ has a time-constant value (but a different one for each point of space), so that

$$I = I_1 + I_2 + 2\sqrt{I_1 I_2}\cos\delta \tag{6.2}$$

At the points of space for which cos $\delta > 0$, the intensity $I$ will exceed $I_1 + I_2$; at the points for which cos $\delta < 0$, it will be smaller than $I_1 + I_2$. Thus, the superposition of coherent light waves is attended by redistribution of the light flux in space. As a result, maxima of the intensity will appear at some spots and minima at others. This phenomenon is called the interference of waves. Interference manifests itself especially clearly when the intensity of both interfering waves is the same: $I_1 = I_2$. Hence, according to Eq. (6.2), at the maxima $I = 4I_1$, while at the minima $I = 0$. For incoherent waves in the same condition, we get the same intensity $I = 2I_1$ everywhere [see Eq. (6.1)].

It follows from what has been said above that when a surface is illuminated by several sources of light (for example, by two lamps), an interference pattern ought to be observed with a characteristic alternation of maxima and minima of intensity. We know from our everyday experience, however, that in this case the illumination of the surface diminishes monotonously with an increasing distance from the light sources, and no interference pattern is observed. The explanation is that natural light sources are not coherent.

The incoherence of natural light sources is due to the fact that the radiation of a luminous body consists of the waves emitted by many atoms.

The individual atoms emit wave trains with a duration of about $10^{-8}$ s and a length of about 3 m. The phase of a new train is not related in any way to that of the preceding one. In the light wave emitted by a body, the radiation of one group of atoms after about $10^{-8}$ s is replaced by the radiation of another group, and the phase of the resultant wave undergoes random changes.

Coherent light waves can be obtained by splitting (by means of reflections or refractions) the wave emitted by a single source into two parts. If these waves are made to cover different optical paths and are then superposed onto each other, interference is observed. The difference between the optical paths covered by the interfering waves must not be very great because the oscillations being added must belong to the same resultant wave train. If this difference will be of the order of one meter, oscillations corresponding to different trains will be superposed, and the phase difference between them will continuously change in a chaotic way.

Assume that the splitting into two coherent waves occurs at point $O$ (Fig. 6.1). Up to point $P$, the first wave travels the path $s_1$ in a medium of refractive index $n_1$, and the second wave travels the path $s_2$ in a medium of refractive index $n_2$. If the phase of oscillations at point $O$ is $\omega t$, then the first wave will produce the oscillation $A_1 \cos \omega \left(t - \frac{s_1}{v_1}\right)$ at point $P$, and the second wave, the oscillation $A_2 \cos \omega \left(t - \frac{s_2}{v_2}\right)$ at this point; $v_1 = c/n_1$ and $v_2 = c/n_2$ are the phase velocities of the waves. Hence, the difference between the phases of the oscillations produced by the waves at point $P$ will be

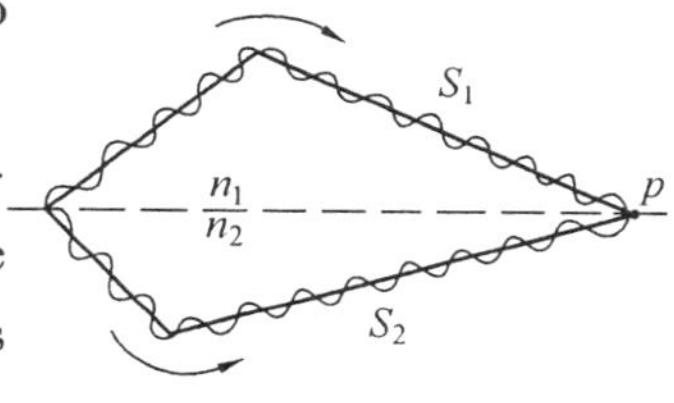


**Fig.6.1**

$$\delta = \omega\left(\frac{s_2}{v_2} - \frac{s_1}{v_1}\right) = \frac{\omega}{c}(n_2 s_2 - n_1 s_1)$$

Replacing $\omega/c$ with $2\pi\nu/c = 2\pi/\lambda_0$ (where $\lambda_0$ is the wavelength in a vacuum), the expression for the phase difference can be written in the form

$$\delta = \frac{2\pi}{\lambda_0}\Delta \tag{6.3}$$

where

$$\Delta = n_2 s_2 - n_1 s_1 = L_2 - L_1 \tag{6.4}$$

is a quantity equal to the difference between the optical paths travelled by the waves and is called the difference in optical path .

A glance at Eq. (6.3) shows that if the difference in the optical path equals an integral number of wavelengths in a vacuum:

$$\Delta = \pm\, m\lambda_0 (m = 0,1,2,\cdots) \tag{6.5}$$

then the phase difference $\delta$ is a multiple of $2\pi$, and the oscillations produced at point $P$ by both waves will occur with the same phase. Thus, Eq. (6.5) is the condition for an interference maximum, i.e. for constructive interference.

If $\Delta$ equals a half-integral number of wavelengths in a vacuum:

$$\Delta = \pm\left(m + \frac{1}{2}\right)\lambda_0 \quad (m = 0,1,2,\cdots) \tag{6.6}$$

then $\delta = \pm(2m+1)\pi$, so that the oscillations at point $P$ are in counterphase. Thus, Eq.(6.6) is the condition for an interference minimum, i.e. for destructive interference.

Let us consider two cylindrical coherent light waves emerging from sources $S_1$ and $S_2$ having the form of parallel thin luminous filaments or narrow slits (Fig.6.2). The region in which these waves overlap is called the interference field. Within this entire region, there are observed alternating places with maximum and minimum intensity of light. If we introduce a screen into the interference field, we shall see on it an interference pattern having the form of alternating light and dark fringes. Let us calculate the width of these fringes, assuming that the screen is parallel to a plane passing through sources $S_1$ and $S_2$. We shall characterize the position of a point on the screen by the coordinate $x$ measured in a direction at right angles to lines $S_1$ and $S_2$. We shall choose the beginning of our readings at point $O$ relative to which $S_1$ and $S_2$ are arranged symmetrically. We shall consider that the sources oscillate in the same phase. Examination of Fig. 6.2 shows that

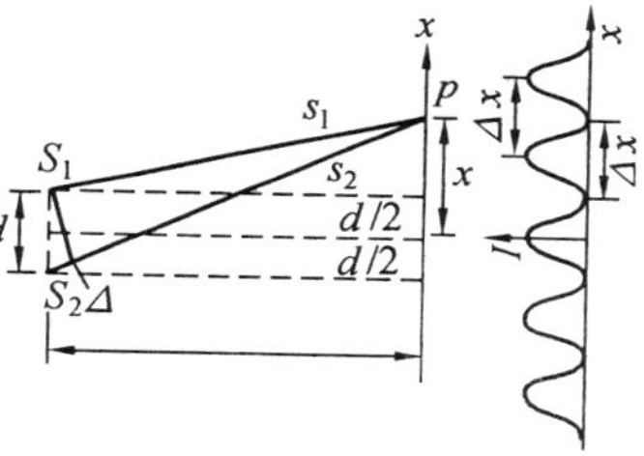


**Fig. 6.2**

$$S_1^2 = l^2 + (x - \frac{d}{2})^2; \quad S_2^2 = l^2 + (x + \frac{d}{2})^2$$

Hence,

$$S_2^2 - S_1^2 = (S_2 + S_1)(S_2 - S_1) = 2xd$$

It will be established somewhat later that to obtain a distinguishable interference pattern, the distance between the sources $d$ must be considerably smaller than the distance to the screen $l$. The distance $x$ within whose limits interference fringes are formed is also considerably smaller than $l$. In these conditions, we can assume that $s_2 + s_1 \approx 2l$. Thus, $s_2 - s_1 = xd/l$. Multiplying $s_2 - s_1$ by the refractive index of the medium $n$, we get the difference in the optical path

$$\Delta = n\frac{xd}{l} \tag{6.7}$$

The introduction of this value of $\Delta$ into condition (6.5) shows that intensity maxima will be observed at values of $x$ equal to

$$x_{\max} = \pm m\frac{l}{d}\lambda \quad (m = 0, 1, 2, 3, \cdots) \tag{6.8}$$

Here $\lambda = \lambda_0/n$ is the wavelength in the medium filling the space between the sources and the screen.

Using the value of $\Delta$ given by Eq. (6.7) in condition (6.6), we get the coordinates of the intensity minima:

$$x_{min} = \pm\left(m + \frac{1}{2}\right)\frac{l}{d}\lambda \quad (m = 0,1,2,\cdots) \tag{6.9}$$

Let us call the distance between two adjacent intensity maxima the distance between interference fringes, and the distance between adjacent intensity minima the width of an interference fringe. It can be seen from Eqs. (6.8) and (6.9) that the distance between fringes. and the width of a fringe have the same value equal to

$$\Delta x = \frac{l}{d}\lambda \tag{6.10}$$

According to Eq. (6.10), the distance between the fringes grows with a decreasing distance $d$ between the sources. If $d$ were comparable with $l$, the distance between the fringes would be of the same order as $\lambda$, i.e. would be several scores of micrometers. In this case the separate fringes would be absolutely indistinguishable. For an interference pattern to become distinct, the above-mentioned condition $d \ll l$ must be observed.

If the intensity of the interfering waves is the same ($I_1 = I_2 = I_0$) then according to Eq.(6.2) the resultant intensity at the points for which the phase difference is $\delta$ is determined by the expression

$$I = 2I_0(1 + \cos\delta) = 4I_0\cos^2\frac{\delta}{2}$$

Since $\delta$ is proportional to $\Delta$ [see Eq. (6.3)], then in accordance with Eq. (6.7) $\delta$ grows proportionally to $x$. Hence, the intensity varies along the screen in accordance with the law of cosine square. The Fig. 6.2 shows the dependence of $I$ on $x$ obtained in monochromatic light.

The width of the interference fringes and their spacing depend on the wavelength $\lambda$. The maxima of all wavelengths will coincide only at the centre a pattern when $x = 0$. With an increasing distance from the center of the pattern, the maxima of different colors become displaced from one another more and more. The result is blurring of the interference pattern when it is observed in white light. The number of distinguishable interference fringes appreciably grows in monochromatic light.

Having measured the distance between the fringes $\Delta x$ and knowing $l$ and $d$, we can use Eq. (6.10) to find $\lambda$. It is exactly from experiments involving the interference of light that the wavelengths for light rays of various colors were determined for the first time.

We have considered the interference of two cylindrical waves. Let us see what happens when two plane waves are superposed. Assume that the amplitudes of these waves are the same, and the directions of their propagation make the angle $2\varphi$ (Fig. 6.3). We shall consider that the directions of oscillations of the light vector are perpendicular to the plane of the drawing. The wave vectors $k_1$ and $k_2$ are in the plane of the drawing and have the same magnitude equal to $k = 2\pi/\lambda$. Let us write the equations of these waves:

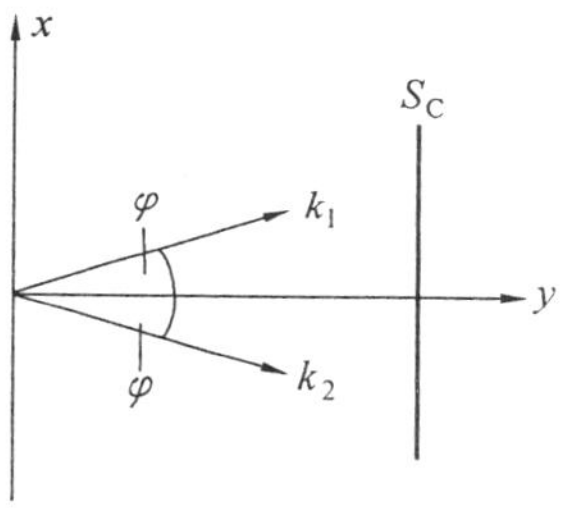


**Fig.6.3**

$$A\cos(\omega t - \boldsymbol{k}_1\boldsymbol{r}) = A\cos(\omega t - k\sin\varphi \cdot x - k\cos\varphi \cdot y)$$

$$A\cos(\omega t - \boldsymbol{k}_2\boldsymbol{r}) = A\cos(\omega t + k\sin\varphi \cdot x - k\cos\varphi \cdot y)$$

The resultant oscillation at points with the coordinates $x$ and $y$ has the form

$$A\cos(\omega t - k\sin\varphi \cdot x - k\cos\varphi \cdot y) + A\cos(\omega t + k\sin\varphi \cdot x - k\cos\varphi \cdot y) = 2A\cos(k\sin\varphi \cdot x)\cos(\omega t - k\cos\varphi \cdot y) \tag{6.11}$$

It follows from this equation that at points where

$$k\sin\psi \cdot x = \pm m\pi \quad (m = 0, 1, 2, \cdots)$$

the amplitude of the oscillations is $2A$; where $k\sin\varphi \cdot x = \pm(m + \frac{1}{2})$

$\pi$, the amplitude of the oscillations is zero. No matter where we place screen *Sc*, which is perpendicular to the $y$-axis, we shall observe on it a system of alternating light and dark fringes parallel to the $z$-axis (this axis is perpendicular to the plane of the drawing). The coordinates of the intensity maxima will be

$$x_{max} = \pm \frac{m\pi}{k\sin\varphi} = \pm \frac{m\lambda}{2\sin\varphi} \qquad (6.12)$$

Only the phase of the oscillations depends on the position of the screen (on the coordinate $y$) [see Eq. (6.11)].

We have assumed for simplicity that the initial phases of interfering waves are zero. If the difference between these phases is other than zero, a constant addend will appear in Eq. (6.12) – the fringe pattern will move along the screen.

## 6.2 COHERENCE

By coherence is meant the coordinated proceeding of several oscillatory or wave processes. The degree of coordination may vary. We can accordingly introduce the concept of the degree of coherence of two waves.

Temporal and spatial coherence are distinguished. We shall begin with a discussion of temporal coherence.

Temporal Coherence. The process of interference described in the preceding section is idealized. This process is actually much more complicated. The reason is that a monochromatic wave described by the expression

$$A\cos(\omega t - kr + \alpha)$$

where $A$, $\omega$, and $\alpha$ are constants, is an abstraction. A real light wave is formed by the superposition of oscillations of all possible frequencies (or wavelengths) confined within a more or less narrow but finite range of frequencies $\Delta\omega$ (or the corresponding range of wavelengths $\Delta\lambda$). Even for light considered to be monochromatic (single-colored), the frequency interval $\Delta\omega$ is finite. In addition, the amplitude of the wave $A$ and the phase $\alpha$ undergo continuous random (chaotic) changes with time. Hence, the oscillations produced at a certain point of space by two superposed light waves have the form

$$A_1(t)\cos[\omega_1(t)\cdot t+\alpha_1(t)];\quad A_2(t)\cos[\omega_2(t)\cdot t+\alpha_2(t)] \tag{6.13}$$

the chaotic changes in the functions $A_1(t)$, $\omega_1(t)$, $\alpha_1(t)$, $A_2(t)$, $\omega_2(t)$, and $\alpha_2(t)$ being absolutely independent.

We shall assume for simplicity's sake that the amplitude $A_1$ and $A_2$ are constant. Changes in the frequency and phase can be reduced either to a change only in the phase, or to a change only in the frequency. Let us write the function

$$f(t)=A\cos[\omega(t)\cdot t+\alpha(t)] \tag{6.14}$$

in the form

$$f(t)=A\cos\{\omega_0 t+[\omega(t)-\omega_0]t+\alpha(t)\}$$

where $\omega_0$ is a certain average value of the frequency, and introduce the notation $[\omega(t)-\omega_0]t+\alpha(t)=\alpha'(t)$. Equation(6.14) will thus become

$$f(t)=A\cos[\omega_0 t+\alpha'(t)] \tag{6.15}$$

we have obtained a function in which only the phase of the oscillation changes chaotically.

On the other hand, it is proved in mathematics that an inharmonic function, for example, function(6.14), can be represented in the form of the sum of harmonic functions with frequencies confined within a certain interval $\Delta\omega$ [see Eq.(6.16)].

Thus, when considering the matter of coherence, two approaches are possible: a "phase" one and a "frequency" one. Let us begin with the phase approach. Assume that the frequencies $\omega_1$ and $\omega_2$ in Eq.(6.13) satisfy the condition $\omega_1=\omega_2=\text{const}$. Now let us find the influence of a change in the phases $\alpha_1$ and $\alpha_2$. According to Eq (6.2), with our assumptions, the intensity of light at a given point is determined by the expression

$$I = I_1 + I_2 + 2\sqrt{I_1 I_2}\cos\delta(t)$$

Where $\delta(t) = \alpha_2(t) - \alpha_1(t)$. The last addend in this equation is called the interference term.

An instrument that can be used to observe an interference pattern (the eye, a photographic plate, etc.) has a certain inertia. In this connection, it registers a pattern averaged over the time interval $t_{instr}$ needed for "operation" of the instrument. If during the time $t_{instr}$ the factor $\cos\delta(t)$ takes on all the values from $-1$ to $+1$, the average value of the interference term will be zero. Therefore the intensity registered by the instrument will equal the sum of he intensities produce at a given point by each of the separatelyinterference is absent, and we are forced to acknowledge that the waves are incoherent.

If during the time $t_{instr}$, however, the value of $\cos\delta(t)$ remains virtually constant, the instrument will detect interference, and the waves must be acknowledged as coherent.

It follows from the above that the concept of coherence is relative: two waves can behave like coherent ones when observed using one instrument (having a low inertia), and like incoherent ones when observed using another instrument (having a high inertia). The coherent properties of waves are characterized by introducing the coherence time $t_{coh}$. It is defined as the time during which a chance change in the wave phase $\alpha(t)$ reaches a value of the order of $\pi$. During the time $t_{coh}$, an oscillation, as it were, forgets its initial phase and becomes incoherent with respect to itself.

Using the concept of the coherence time, we can say that when the instrument time is much greater than the coherence time of the superposed waves ($t_{instr} \gg t_{coh}$), the instrument does not register interference, when

($t_{instr} \gg t_{coh}$) the instrument will detect a sharp interference pattern. At intermediate values of $t_{instr}$, the sharpness of the pattern will diminish as $t_{instr}$ grows from values smaller than $t_{coh}$ to values greater than it.

The distance $l_{coh} = ct_{coh}$ over which a wave travels during the time $t_{coh}$ is called the coherence length (or the train length). The coherence length is the distance over which a chance change in the phase reaches a value of about $\pi$. To obtain an interference pattern by splitting a natural wave into two parts, it is essential that the optical path difference $\Delta$ be smaller than the coherence length. This requirement limits the number of visible interference fringes observed when using the layout shown in Fig. 6.2. An increase in the fringe number m is attended by a growth in the path difference. As a result, the sharpness of the fringes becomes poorer and poorer.

Let us pass over to a consideration of the part of the non-monochromatic nature of light waves. Assume that light consists of a sequence of identical trains of frequency $\omega_0$ and duration $\tau$. When one train is replaced with another one, the phase experiences disordered changes. As a result, the trains are mutually incoherent. With these assumptions, the duration of a train $\tau$ virtually coincides with the coherence time $t_{coh}$.

In mathematics, the Fourier theorem is proved, according to which any finite and integrable function $F(t)$ can be represented in the form of the sum of an infinite number of harmonic components with a continuously changing frequency:

$$F(t) = \int_{-\infty}^{+\infty} A(\omega)e^{i\omega t}d\omega \tag{6.16}$$

Expression (6.16) is known as the Fourier integral. The function $A(\omega)$ inside the integral is the amplitude of the relevant monochromatic component. According to the theory of Fourier integrals, the analytical form of the function $A(\omega)$ is determined by the expression

$$A(\omega) = 2\pi\int_{-\infty}^{+\infty} F(\xi)\mathrm{e}^{-i\omega\xi}\mathrm{d}\xi \tag{6.17}$$

where $\xi$ is an auxiliary integration variable.

Assume that the function $F(t)$ describes a light disturbance at a certain point at the moment of time $t$ due to a single wave train. Hence, it is determined by the conditions

$$F(t) = A_0\exp(i\omega_0 t) \qquad aat \mid t \mid \leqslant \frac{\tau}{2}$$

$$F(t) = 0 \qquad aat \mid t \mid > \frac{\tau}{2}$$

A graph of the real part of this function is given in Fig. 6.4.

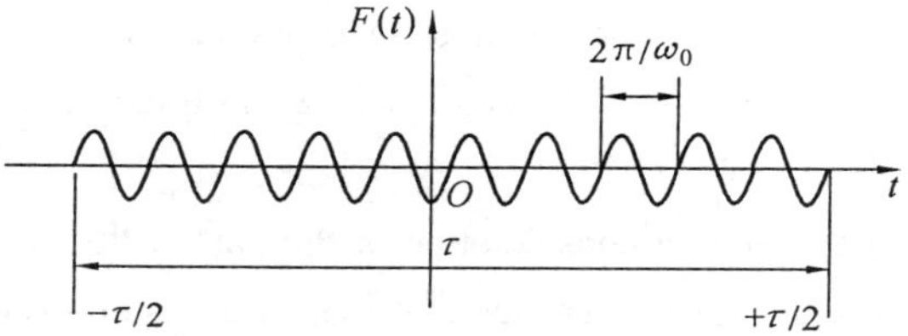


**Fig.6.4**

Outside the interval from $-\tau/2$ to $+\tau/2$, the function $F(t)$ is zero. Therefore, expression (6.17) determining the amplitude of the harmonic components has the form

$$A(\omega) = 2\pi\int_{-\tau/2}^{+\tau/2}[A_0\exp(i\omega_0\xi)]\exp(-i\omega\xi)\mathrm{d}\xi =$$

$$2\pi A_0\int_{-\tau/2}^{+\tau/2}[\exp i(\omega_0-\omega)\xi]\mathrm{d}\xi =$$

$$2\pi A_0\frac{\exp[i(\omega_0-\omega)\xi]}{i(\omega_0-\omega)}\Bigg|_{+\tau/2}^{-\tau/2}$$

After introducing the integration limits and simple transformations, we arrive at the equation

$$A(\omega) = \pi A_0\tau\frac{\sin[(\omega_0-\omega)\tau/2]}{[(\omega-\omega_0)\tau/2]}$$

The intensity $I(\omega)$ of a harmonic wave component is proportional to the square of the amplitude, i.e. to the expression

$$f(\omega) = \frac{\sin^2[(\omega_0 - \omega)\tau/2]}{[(\omega - \omega_0)\tau/2]^2} \tag{6.18}$$

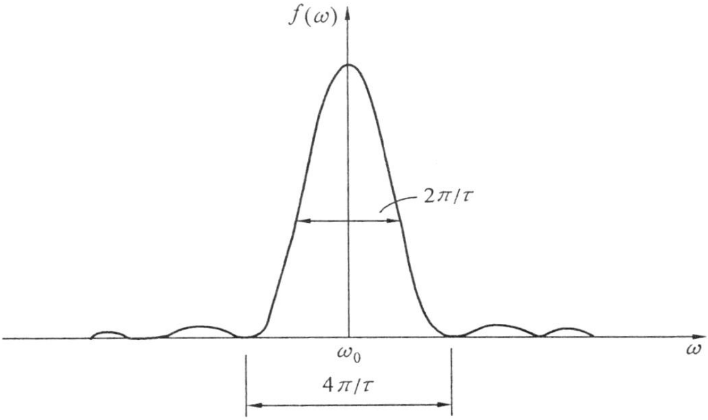


**Fig.6.5**

A graph of function (6.18) is shown in Fig.6.5. A glance at the figure shows that the intensity of the components whose frequencies are within the interval of width $\Delta\omega = 2\pi/\tau$ considerably exceeds the intensity of the remaining components. This circumstance allows us to relate the duration of a train $\tau$ to the effective frequency range $\Delta\omega$ of a Fourier spectrum:

$$\tau = \frac{2\pi}{\Delta\omega} = \frac{1}{\Delta\nu}$$

Identifying $\tau$ with the coherence time, we arrive at the relation

$$t_{coh} \sim \frac{1}{\Delta\nu} \tag{6.19}$$

(The sign ~ stands for "equal to in the order of magnitude")

It can be seen from expression (6.19) that the broader the interval of frequencies present in a given light wave, the smaller is the coherence time of this wave.

The frequency is related to the wavelength in a vacuum by the expressly $\nu = c/\lambda_0$. Differentiation of this expression yields $\Delta\nu = c\Delta\lambda_0/\lambda_0^2 \approx c\Delta\lambda/\lambda^2$ (we have omitted the minus sign obtained in differentiation and also assumed that $\lambda_0 \approx \lambda$). Substituting for $\Delta\nu$ in E.q. (6.19) its expression through $\lambda$ and $\Delta\lambda$, we obtain the following expression for the coherence time:

$$t_{coh} \sim \frac{\lambda^2}{c\Delta\lambda} \tag{6.20}$$

Hence, we get the following value for the coherence length:

$$l_{coh} = ct_{coh} \sim \frac{\lambda^2}{\Delta\lambda} \tag{6.21}$$

Examination of Eq.(6.5) shows that the path difference at which a maximum of the $m$-th order is obtained is determined by the relation

$$\Delta_m = \pm m\lambda_0 \approx \pm m\lambda$$

When this path difference reaches values of the order of the coherence-length. the fringes become indistinguishable. Consequently, the extreme interference order observed is determined by the condition

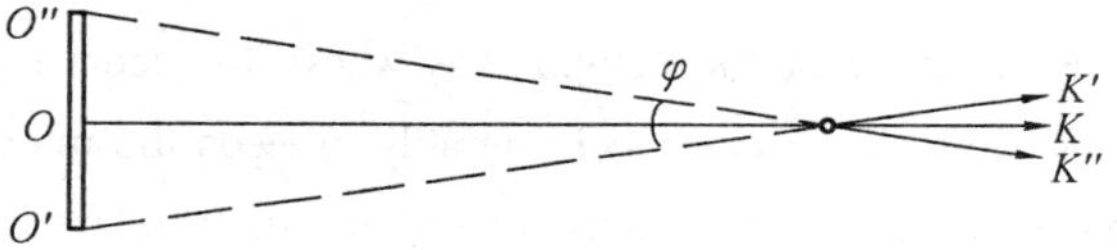


**Fig.6.6**

$$m_{extr}\lambda \sim l_{coh} \sim \frac{\lambda^2}{\Delta\lambda}$$

whence

$$m_{extr} \sim \frac{\lambda}{\Delta\lambda} \tag{6.22}$$

It follows from Eq. (6.22) that the number of interference fringes observed according to the layout shown in Fig. 6.2 grows when the wavelength interval in the light used diminishes.

Spatial Coherence. According to the equation $k = \omega/v = n\omega/c$, scattering of the frequencies $\Delta\omega$ results in scattering of the values of $k$. We have established that the temporal coherence is determined by the value of $\Delta\omega$. Consequently, the temporal coherence is associated with scattering of the values of the magnitude of the wave vector $\boldsymbol{k}$. Spatial coherence is associated with scattering of the directions of the vector $\boldsymbol{k}$ that is characterized by the quantity $\Delta\boldsymbol{e}_k$.

The setting up at a certain point of space of oscillations produced by waves with different values of $\boldsymbol{e}_k$ is possible if these waves are emitted by different sections of an extended (not a point) light source. Let us assume for simplicity's sake that the source has the form of a disk visible from a given point at the angle $\varphi$. It can be seen from Fig.6.6 that the angle $\varphi$ characterizes the interval confining the unit vectors $\boldsymbol{e}_k$. We shall consider that this angle is small.

Assume that the light from the source falls on two narrow slits behind which there is a screen (Fig.6.7). We shall consider that the interval of frequencies emitted by the source is very small. This is needed for the degree of temporal coherence to be sufficient for obtaining a sharp interference pattern. The wave arriving from the section of the surface designated in Fig.6.7 by $O$ produces a zero-order maximum $M$ at the middle of the screen. The zero-order maximum $M'$ produced by the wave arriving

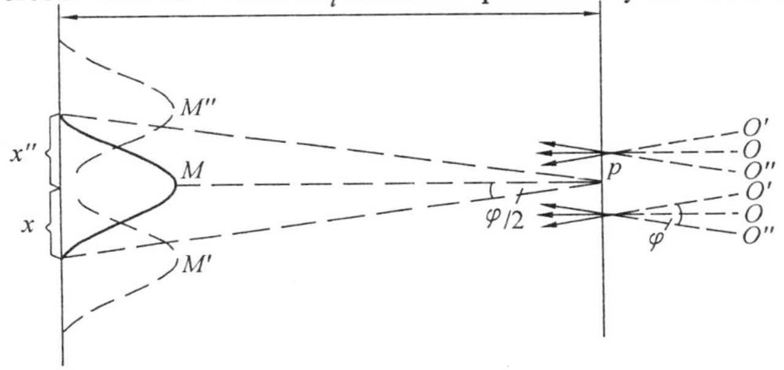


**Fig.6.7**

*from section* $O'$ will be displaced from the middle of the screen by the distance $x'$. Owing to the smallness of the angle $\varphi$ and of the ratio $d/l$, we can consider that $x' = l\varphi/2$.

The zero-order maximum $M''$ produced by the wave arriving from section $O''$ is displaced in the opposite direction from the middle of the screen over the distance $x''$ equal to $x'$. The zero-order maxima from the other sections of the source will be between the maxima $M'$ and $M''$.

The separate sections of the light source produce waves whose phases are in no way related to one another. For this reason, the interference pattern appearing on the screen will be a superposition of the patterns produced by each section separately. If the displacement $x'$ is much smaller than flip width of an interference fringe $\Delta x = l\lambda/d$ [see Eq. (6.10)], then the maxima from different sections of the source will practically be superposed on one another, and the pattern will be like one produced by a point source. When $x' \approx \Delta x$, the maxima from some sections will coincide with the minima from others, and no interference pattern will be observed. Thus, an interference pattern will be distinguishable provided that $x' < \Delta x$, i.e.

$$\frac{l\varphi}{2} < \frac{l\lambda}{d} \tag{6.23}$$

or

$$\varphi < \frac{\lambda}{d} \tag{6.24}$$

We have omitted the factor 2 when passing over from expression (6.23) to (6.24).

Formula (6.24) determines the angular dimensions of a source at which interference is observed. We can also use this formula to find the greatest distance between the slits at which interference from a source with the angular dimension $\varphi$ can still be observed. Multiplying inequality (6.24) by $d/\varphi$ we arrive at the condition

$$d < \frac{\lambda}{\varphi} \tag{6.25}$$

A collection of waves with different values of $e_k$ can be replaced with the resultant wave falling on a screen with slits. The absence of an interference pattern signifies that the oscillations produced by this wave at the places where the first and second slits are situated are incoherent. Consequently, the oscillations in the wave itself at points at a distance $d$ apart are incoherent, too. If the source were ideally monochromatic (this means that $\Delta\nu = 0$ and $t_{coh} = \infty$), the surface passing through the slits would be a wave one, and the oscillations at all the points of this surface would occur in the same phase. We have established that when $\Delta\nu \neq 0$ and the source has finite dimensions ($\varphi \neq 0$), the oscillations at points of a surface at a distance of $d > \lambda/\varphi$ are incoherent.

We shall call a surface which would be a wave one if the source were monochromatic a pseudowave surface for brevity. We could satisfy condition (6.24) by reducing the distance $d$ between the slits, i.e. by taking closer points of the pseudowave surface. Consequently, oscillations produced by a wave at adequately close points of a pseudowave surface are coherent. Such coherence is called spatial.

Thus, the phase of an oscillation changes chaotically when passing from one point of a pseudowave surface to another. Let us introduce the distance $\rho_{coh}$, upon displacement by which along a pseudowave surface a random change in the phase reaches a value of about $\pi$. Oscillations at two points of a pseudowave surface spaced apart at a distance less than $\rho_{coh}$ will be approximately coherent. The distance $\rho_{coh}$ is called the spatial coherence length or the coherence radius. It can be seen from expression (6.25) that

$$\rho_{coh} \sim \frac{\lambda}{\varphi} \tag{6.26}$$

The angular dimension of the Sun is about 0.01 radian, and the length of its light waves is about 0.5 $\mu$m. Hence, the coherence radius ot the light waves arriving from the Sun has a value of the order of

$$\rho_{coh} = \frac{0.5}{0.01} = 50\ \mu m = 0.05\ mm \qquad (6.27)$$

The entire space occupied by a wave can be divided into parts in each of which the wave approximately retains coherence. The volume of such a part of space, called the coherence volume, in its order of magnitude e-quals the product of the temporal coherence length and the area of a circle of radius $\rho_{coh}$.

The spatial coherence of a light wave near the surface of the heated body emitting it is restricted by a value of $\rho_{coh}$ of only a few wavelengths. With an increasing distance from the source, the degree of spatial coherence grows. The radiation of a laser has an enormous temporal and spatial coherence. At the outlet opening of a laser, spatial coherence is observed throughout the entire cross section of the light beam.

It would seem possible to observe interference by passing light propagating from an arbitrary source through two slits in an opaque screen. With a small spatial coherence of the wave falling on the slits, however, the beams of light passing through them will be incoherent, and an interference pattern will be absent. The English scientist Thomas Young (1772 ~ 1829) in 1802 obtained interference from two slits by increasing the spatial coherence of the light falling on the slits. Young achieved such an increase by first passing the light through a small aperture in an opaque screen. This light was used to illuminate the slits in a second opaque screen. Thus, for the first time in history, Young observed the interference of light waves and determined the lengths of these waves.

## 6.3 WAYS OF OBSERVING THE INTERFERENCE OF LIGHT

Let us consider two concrete interference layouts of which one uses reflection for splitting a light wave into two parts, and the other refraction of light.

**Fresnel's Double Mirror.** Two plane contacting mirrors $OM$ and $ON$ are arranged so that their reflecting surfaces form an obtuse angle close to $\pi$(Fig.6.8). Hence, the angle $\varphi$ in the figure is very small. A straight light source $S$ (for example, a narrow luminous slit) is placed parallel to the line of intersection of the mirrors $O$ (perpendicular to the plane of the drawing) at a distance $r$ from it. The mirrors reflect two cylindrical coherent waves onto screen $Sc$. They propagate as if they were emitted by virtual sources $S_1$ and $S_2$, Opaque screen $S_{C1}$ prevents the direct propagation of the light from source $S$ to screen $S_c$.

Ray $OQ$ is the reflection of ray $SO$ from mirror $OM$, and ray $OP$ is the reflection of ray $SO$ from mirror $ON$. It is easy to see that the angle between rays $OP$ and $OQ$ is $2\varphi$. Since $S$ and $S_1$, are symmetrical relative to $OM$, the length of segment $OS_1$ equals $OS$, i.e. $r$. Similar reasoning leads to the same result for segment $OS_2$. Thus, the distance between sources $S_1$ and $S_2$ is

$$d = 2r\sin\varphi \approx 2r\varphi$$

Inspection of Fig.6.8 shows that $a = r\cos\varphi \approx r$. Hence,

$$l = r + b$$

where $b$ is the distance from the line of intersection of the mirrors $O$ to screen $Sc$.

Using the values of $d$ and $l$ we have found in Eq. (6.10), we obtain the width of an interference fringe

$$\Delta x = \frac{r + b}{2r\varphi}\lambda \tag{6.28}$$

The region of wave overlapping $PQ$ has a length of $2b\tan\varphi \approx 2b\varphi$. Dividing this length by the width of a fringe $\Delta x$, we find the maximum number of interference fringes that can be observed with the aid of Fresnel's double mirror at the given parameters of a layout:

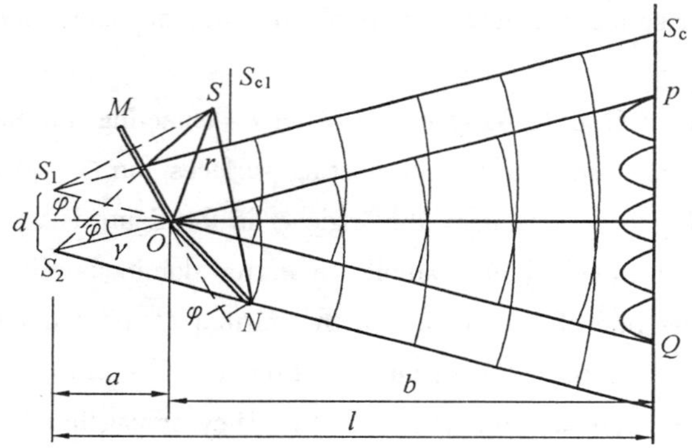


**Fig.6.8**

$$N = \frac{4br\varphi^2}{\lambda(r+b)} \tag{6.29}$$

For all these fringes to be visible indeed, it is essential that $N/2$ be not greater than $m_{extr}$ determined by expression (6.22)

**Fresnel's Biprism.** Two prisms with a small refracting angle $\theta$ made from a single piece of glass have one common face (Fig.6.9). A straight light source $S$ is arranged parallel to this face at a distance $a$ from it.

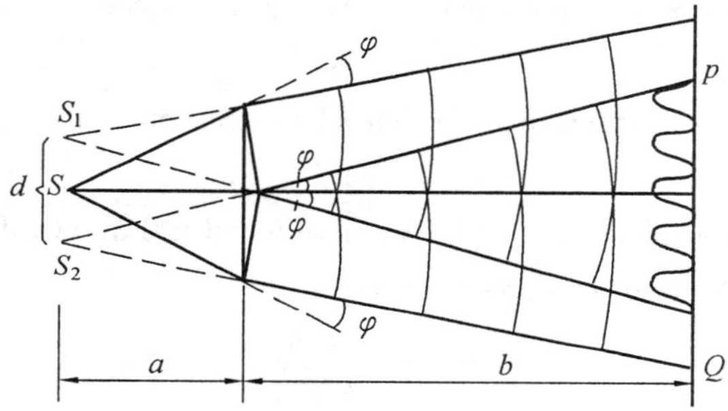


**Fig.6.9**

It can be shown that when the refracting angle $\theta$ of the prism is very small and the angles of incidence of the rays on the face of the prism are not very great, all the rays are deflected by the prism through a practically identical angle equal to

$$\varphi = (n - 1)\theta$$

($n$ is the refractive index of the prism). The angle of incidence of the rays on the biprism is not great. Therefore, all the rays are deflected by each half of the biprism through the same angle. As a result, two coherent cylindrical waves are formed emerging from virtual sources $S_1$ and $S_2$ in the same plane as $S$. The distance between the sources is

$$d = 2a\sin\varphi \approx 2a\varphi = 2a(n - 1)\theta$$

The distance from the sources to the screen is

$$l = a + b$$

We find the width of an interference fringe by Eq. (6.10):

$$\Delta x = \frac{a + b}{(n - 1)\theta}\lambda \tag{6.30}$$

The region of overlapping of the waves $PQ$ has the length

$$2b\tan\varphi \approx 2b\varphi = 2b(n - 1)\theta$$

The maximum number of fringes observed is

$$N = \frac{4ab(n - 1)^2\theta^2}{\lambda(a + b)} \tag{6.31}$$

## The Further Study:

### The Michelson Interferometer

Many varieties of interference instruments called interferometers are in use Figure 6.10 is a schematic view of a Michelson interferometer. A light beam from source $S$ falls on semitransparent plate $P_1$ coated with a thin layer of silver (this layer is depicted by dots in the figure). Half of the incident light flux is reflected by plate $P_1$ in the direction of ray 1, and half passes through the plate and propagates in the direction of ray 2. Beam 1 is reflected from mirror $M_1$ and returns to $P_1$, where it is split into two beams of equal intensity. One of them passes through the plate and forms beam 1′, and the second one is reflected in the direction of $S$. The latter beam will no longer interest us. Beam 2 after being reflected by mirror $M_2$ also returns to plate $P_1$ where it is divided into two parts: beam 2′ reflected from the semitransparent layer, and the beam transmitted through the layer, which will also no longer interest us. Light beams 1′ and 2′ have the same intensity.

If conditions of temporal and spatial coherence are observed, beams 1′ and 2′ will interfere. The result of this interference depends on the optical path difference from plate $P_1$ to mirrors $M_1$ and $M_2$ and back. Ray 2 passes thrush the plate three times, and ray 1 only once. To compensate the resulting change in the optical path difference (owing to dispersion) for waves of different lengths, plate $P_2$ is placed in the path of ray 1. Plates $P_1$ and $P_2$ are identical, except for the silver coating on the former. This arrangement makes the paths of rays 1 and 2 in glass equal. The interference pattern is observed with the aid of telescope $T$.

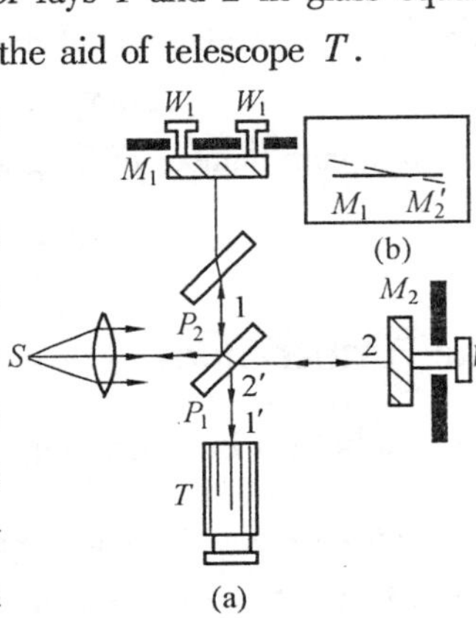


Fig.6.10

Let us mentally replace mirror $M_2$ with its virtual image $M'_2$ in semitransparent plate $P_1$. Beams 1′ and 2′ can thus be considered as due to reflection from a transparent plate contained between planes $M_2$ and $M'_2$. We can use adjusting screws $W_1$ to change the angle between these planes; in particular, they can be arranged strictly parallel to each other. By rotating micrometric screw $W_2$, we can smoothly move mirror $W_2$ without changing its inclination. We can thus change the thickness of the "plate"; in particular, we can make planes $M_1$ and $M'_2$ intersect (Fig. 6.10(b)).

The nature of the interference pattern depends on the adjustment of the mirrors and on the divergence of the beam of light falling on the instrument. If the beam is parallel, and planes $M_1$ and $M'_2$, make an angle other than zero, then straight fringes of equal thickness parallel to the lines of intersection of planes $M_1$ and $M'_2$ will be observed m the field of vision of the telescope. In white light, all the fringes except the one coinciding with the line of intersection of the zero-order fringe will be colored. The zero-order fringe will be black because beam 1 is reflected from plate $P_1$ from the outside, and beam 2 from the inside. As a result, a phase difference equal to $\pi$ is produced between them. In white light, fringes are observed only with a small thickness of "plate" $M_1M'_2$. In monochromatic light corresponding to the red line of cadmium, Michelson observed a distinct interference pattern at a path difference of the order of 500 000 wavelengths (the distance between $M_1$ and $M_2$ in this case is about 150 mm).

With a slightly diverging beam of light and a strictly parallel arrangement of planes $M_1$ and $M'_2$, fringes of equal inclination are obtained that have the form of concentric rings. When micrometric screw $W_2$ is rotated, the diameter of the rings grows or diminishes. Either new rings appear at the center of the pattern, or the diminishing rings shrink to a point and then vanish. Displacement of the pattern by one fringe corresponds to movement of mirror $M_2$ through half a wavelength.

Michelson used the instrument described above to carry out several experiments that entered the annals of physics. The most famous of them, performed together with the American chemist Edward Morley (1838 ~ 1923) in 1887, had the aim of detecting motion of the Earth relative to the hypothetic ether. In 1890 ~ 1895, Michelson used the interferometer he had invented to make the first comparison of the wavelength of the red line of cadmium with the length of the standard meter.

In 1920, Michelson constructed a stellar interferometer which he used to measure the angular dimensions of stars. This instrument was mounted on a telescope. A screen with two slits was installed in front of the objective of the telescope (Fig. 6.11). The light from a star was reflected from a symmetrical system of mirrors $M_1$, $M_2$, $M_3$ and $M_4$ installed on a rigid frame fastened on a carriage The inner mirrors $M_3$ and $M_4$, were fixed, and the outer ones $M_1$ and $M_2$ could move symmetrically away from or toward mirrors $M_3$ and $M_4$. The path of the rays is clear from the figure. Interference fringes were produced in the focal plane of the telescope objective. Their visibility depended on the distance between the outer mirrors. By moving these mirrors, Michelson determined the distance $l$ between them at which the visibility of the fringes vanishes.

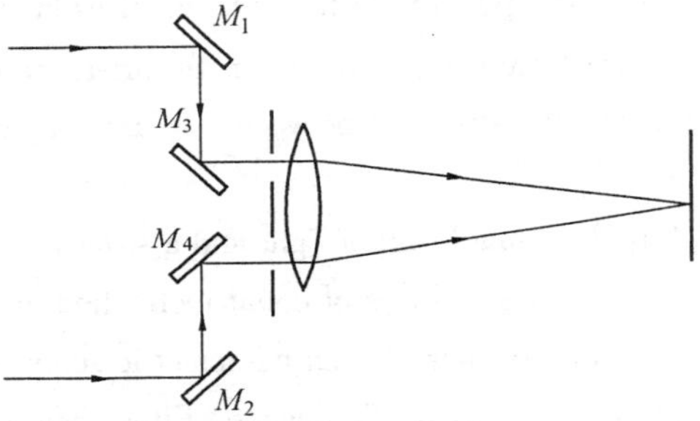


**Fig.6.11**

This distance must be of the order of the coherence radius of a light wave arriving from a star. According to expression (6.26), the coherence radius is $\lambda/\varphi$. The condition $l = \lambda/\varphi$ gives the angular diameter of a star

$$\varphi = \frac{\lambda}{l}$$

Accurate calculations give the formula

$$\varphi = A\frac{\lambda}{l}$$

where $A = 1.22$ for a source in the form of a uniformly illuminated disk. If the disk is darker at its edges than at the centre, the coefficient exceeds 1.22, its value depending on the rate of diminishing of the illumination in the direction from the centre toward the edge. In addition, accurate calculations show that after vanishing at a certain value of $l$, the visibility upon a further increase in $l$ again becomes other than zero; however, the values it reaches are not great.

The maximum distance between the outer mirrors in the stellar interferometer constructed by Michelson was 6.1 m (the diameter of the telescope was 2.5 m). A minimum measurable angular diameter of about 0.02″ corresponded to this distance. The first star whose angular diameter was measured was Betelgeuse (alpha Orion). The value of $\varphi$ obtained for it was 0.047″.

## Summary of Key Terms

**Huygens' principle** The theory by which light waves spreading out from a point source can be regarded as the superposition of tiny secondary wavelets.

**Interference** The superposition of waves producing regions of reinforcement and regions of cancellation. Constructive interference refers to regions of reinforcement; destructive interference refers to regions of cancellation. The interference of selected wavelengths of light produces colors known as interference colors.

7

# Diffraction of Light

## 7.1 INTRODUCTION

By diffraction is meant the combination of phenomena observed when light propagates in a medium with sharp heterogeneities and associated with deviations from the laws of geometrical optics. Diffraction, in particular, leads to light waves bending around obstacles and to the penetration of light into the region of a geometrical shadow. The bending of sound waves around obstacles (i.e. the diffraction of sound waves) is constantly observed in our everyday life. To observe the diffraction of light waves, special conditions must be set up. This is due to the smallness of the lengths of light waves. We know that in the limit, when $\lambda \to 0$, the laws of wave optics transform into those of geometrical optics. Hence, other conditions being equal, the deviations from the laws of geometrical optics decrease with a diminishing wavelength.

There is no appreciable physical difference between interference and diffraction. Both phenomena consist in the redistribution of the light flux as a result of superposition of the waves. For historical reasons, the redistribution of the intensity produced as a result of the superposition of waves emitted by a finite number of discrete coherent sources has been called the interference of waves. The redistribution of the intensity produced as a result of the superposition of waves emitted by coherent sources arranged continuously has been called the diffraction of waves. We therefore speak about the interference pattern from two narrow slits and about the diffraction pattern from one slit.

Diffraction is usually observed by means of the following set-up. An opaque barrier closing part of the wave surface of the light wave is placed in the path of a light wave propagating from a certain source. A screen on which the diffraction pattern appears is placed after the barrier.

Two kinds of diffraction are distinguished. If the light source $S$ and the point of observation $P$ are so far from a barrier that the rays falling on the barrier and those travelling to point $P$ form virtually parallel beams, we have to do with diffraction in parallel rays or with Fraunhofer diffraction. Otherwise, we have to do with Fresnel diffraction. Fraunhofer diffraction can be observed by placing a lens after light source $S$ and another one in front of point of observation $P$ so that points $S$ and $P$ lie in the focal plane of the relevant lens. (Fig. 7.1)

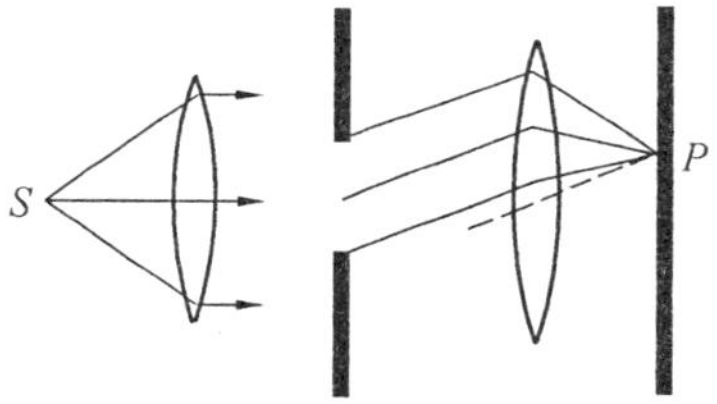


**Fig. 7.1**

## 7.2 HUYGENS-FRESNEL PRINCIPLE

The penetration of light waves into the region of a geometrical shadow can be explained with the aid of Huygens' principle. This principle, however, gives no information on the amplitude and, consequently, on the intensity of waves propagating in different directions. The French physicist Augustin Fresnel (1788 ~ 1827) supplemented Huygens' principle with the concept of the interference of secondary waves. Taking into account the amplitudes and phases of the secondary waves makes it possible to find the amplitude of the resultant wave for any point of space. Huygens' principle developed in this way was named the Huygens – Fresnel principle.

According to the Huygens – Fresnel principle. Every element of wave surface $S$ (Fig.7.2) is the source of a secondary spherical wave whose amplitude is proportional to the size of element $\mathrm{d}S$. The amplitude of a spherical wave diminishes with the distance $r$ from its source according to the law $1/r$ . Consequently, the oscillation

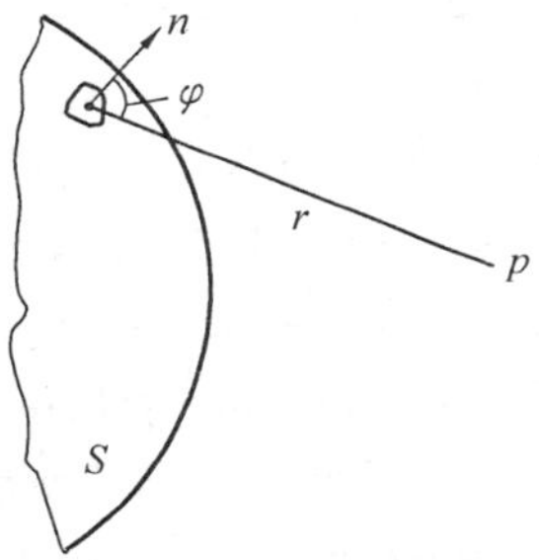


**Fig.7.2**

$$\mathrm{d}E = K\frac{a_0\mathrm{d}s}{r}\cos(\omega t - kr + \alpha_0) \tag{7.1}$$

arrives from each section $\mathrm{d}S$ of a wave surface at point $P$ in front of this surface. In Eq. (7.9), $\omega t + \alpha_0$ is the phase of the oscillation where wave surface $S$ is, $k$ is the wave number, $r$ is the distance from surface element $\mathrm{d}S$ to point $P$. The factor $\alpha_0$ is determined by the amplitude of the light oscillation at the location of $\mathrm{d}S$. The coefficient $K$ depends on the angle $\varphi$ between a normal n to area $\mathrm{d}S$ and the direction from $\mathrm{d}S$ to point $P$. When $\varphi = 0$, this coefficient is maximum; when $\varphi = \frac{\pi}{2}$, it vanishes.

The resultant oscillation at point $P$ is the superposition of the oscillations given by Eq. (7.1) taken for the entire wave surface $S$:

$$E = \int_S K(\varphi)\frac{\alpha_0}{r}\cos(\omega t - kr + \alpha_0)\mathrm{d}S \tag{7.2}$$

This equation is an analytical expression of the Huygens – Fresnel principle.

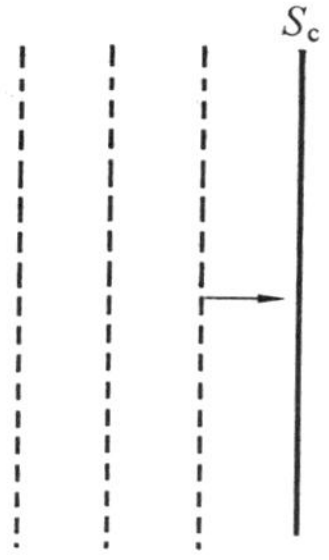

**Fig.7.3**

The Huygens – Fresnel principle can be substantiated by the following reasoning. Assume that thin opaque screen *Sc* (Fig.7.3) is placed in the path of a light wave (we shall consider it plane for simplicity's sake). The intensity of the light everywhere after the screen will be zero. The reason is that the light wave falling on the screen produces oscillations of the electrons in the material of the screen. The oscillating electrons emit electromagnetic waves. The field after the screen is a superposition of the primary wave (falling on the screen) and all the secondary waves. The amplitudes and phases of the secondary waves are such that upon superposition of these waves with the primary one, a zero amplitude is obtained at any point $P$ after the screen. Consequently, if the primary wave produces the oscillation

$$A_{\text{prim}}\cos(\omega t + \alpha)$$

at point $P$, then the resultant oscillation produced by the secondary waves at the same point has the form

$$A_{\text{sec}}\cos(\omega t + \alpha - \pi)$$

Here $A_{\text{sec}} = A_{\text{prim}}$.

What has been said above signifies that when calculating the amplitude of an oscillation set up at point $P$ by a light wave propagating from a real source, we can replace this source with a collection of secondary sources arranged along the wave surface. This is exactly the essence of Huygens – Fresnel principle.

Let us divide the opaque barrier into two parts. One of them, which we shall call a stopper, has finite dimensions and an arbitrary shape (a circle, rectangle, etc.). The other part includes the entire remaining surface of the infinite barrier. As long as the stopper is in place, the resultant oscillation at point $P$ after the barrier is zero. It can be represented as the sum of the oscillations set up by the primary wave, the wave produced by the stopper, and the wave produced by the remaining part of the barrier:

$$A_{\text{prim}}\cos(\omega t + \alpha) + A_{\text{stop}}\cos(\omega t + \alpha') + A_{\text{bar}}\cos(\omega t + \alpha'') = 0 \tag{7.3}$$

If the stopper is removed, i.e. the wave is transmitted through the apreture in the opaque barrier, then the oscillation at point $P$ will have the form

$$E_p = A_{\text{prim}}\cos(\omega t + \alpha) + A_{\text{bar}}\cos(\omega t + \alpha'') = - A_{\text{stop}}\cos(\omega t + \alpha') = A_{\text{stop}}\cos(\omega t + \alpha' - \pi)$$

We have used condition (7.3) and assumed that removal of the stopper does not change the nature of the oscillations of the electrons in the remaining part of the barrier.

We can thus consider that the oscillations at point $P$ are produced by a collection of sources of secondary waves on the surface of the aperture formed after removal of the stopper.

## 7.3 FRESNEL ZONES

The performance of calculations by Eq. (7.2) is a very difficult task in the general case. As Fresnel showed, however, the amplitude of the resultant oscillation can be found by simple algebraic or geometrical summation in cases distinguished by symmetry.

To understand the essence of the method developed by Fresnel, let us determine the amplitude of the light oscillation set up at point $P$ by a spherical wave propagating in an isotropic homogeneous medium from point source $S$ (Fig.7.4). The wave surfaces of such waves are symmetrical relative to straight line $SP$. Taking advantage of this circumstance, let us divide the wave surface shown in the figure into annular zones constructed so that the distances from the edges of each zone to point $P$ differ by $\lambda/2$ ($\lambda$ is the length of the wave in the medium in which it is propagating). Zones having this property are known as Fresnel zones.

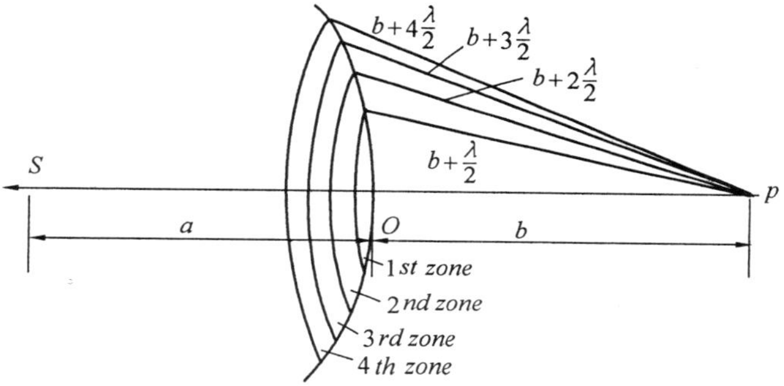


**Fig.7.4**

A glance at Fig.7.4 shows that the distance $b_m$ from the outer edge of the $m$-th zone to point $P$ is

$$b_m = b + m\frac{\lambda}{2} \tag{7.4}$$

($b$ is the distance from the crest $O$ of the wave surface to point $P$).

The oscillations arriving at point $P$ from similar points of two adjacent zones (i.e. from points at the middle of the zones, or at the outer edges of the zones, etc.) are in counterphase. Therefore, the resultant oscillations produced by each of the zones as a whole will differ in phase for adjacent zones by $\pi$ too.

Let us calculate the areas of the zones. The outer boundary of the $m$-th zone separates a spherical segment of height $h_m$ on the wave surface (Fig.7.5). Let the area of this segment be $S_m$. Hence, the area of the $m$-th zone can be written as

$$\Delta S_m = S_m - S_{m-1}$$

where $S_{m-1}$ is the area of the spherical segment separated by the outer boundary of the $(m-1)$-th zone. It can be seen from Fig.7.5 that

$$r_m^2 = a^2 - (a - h_m)^2 = (b + m\frac{\lambda}{2})^2 - (b + h_m)^2$$

where $a$ = radius of the wave surface

$r_m$ = radius of the outer boundary of the $m$-th zone.

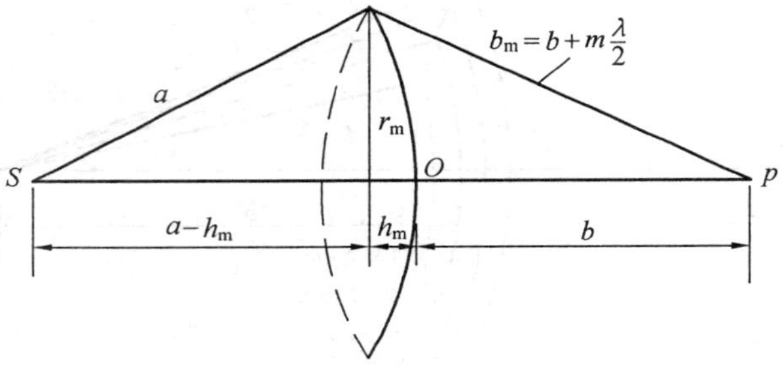


**Fig.7.5**

Squaring the terms in parentheses, we get

$$r_m^2 = 2ah_m - h_m^2 = bm\lambda + m^2\left(\frac{\lambda}{2}\right)^2 - 2bh_m - h_m^2 \quad (7.5)$$

whence

$$h_m = \frac{bm\lambda + m^2(\lambda/2)^2}{2(a+b)} \quad (7.6)$$

Restricting ourselves to a consideration of not too great m's, we may disregard the addend containing $\lambda^2$ owing to the smallness of $\lambda$. In this approximation

$$h_m = \frac{bm\lambda}{2(a+b)} \quad (7.7)$$

The area of a spherical segment is $S = 2\pi Rh$ (here $R$ is the radius of the sphere and $h$ is the height of the segment). Hence,

$$S_m = 2\pi a h_m = \frac{\pi ab}{a+b} m\lambda$$

and the area of the $m$ – th zone is

$$\Delta S_m = S_m - S_{m-1} = \frac{\pi ab\lambda}{a+b}$$

The expression we have obtained does not depend on $m$. This signifies that when $m$ is not too great, the areas of the Fresnel zones are approximately identical.

We can find the radii of the zones from Eq. (7.5). When m is not too great, the height of a segment $h_m \ll a$, and we can therefore consider that $r_m^2 = 2ah_m$. Substituting for $h_m$ its value from Eq. (7.7), we get the following expression for the radius of the outer boundary of the $m$ - th. zone:

$$r_m = \sqrt{\frac{ab}{a+b}m\lambda} \tag{7.8}$$

If we assume that $a = b = 1$ m and $\lambda = 0.5$ $\mu$m, then we get a value of $r_1 = 0.5$ mm for the radius of the first (central) zone. The radii of the following zones grow as $\sqrt{m}$.

Thus, the areas of the Fresnel zones are approximately the same. The distance $b_m$ from a zone to point $P$ slowly increases with the zone number $m$. The angle $\varphi$ between a normal to the zone elements and the direction toward point $P$ also grows with $m$. All this leads to the fact that the amplitude $A_m$ of the oscillation produced by the $m$-th zone at point $P$ diminishes monotonously with increasing $m$. Even at very high values of $m$, when the area of a zone begins to grow appreciably with $m$ [see Eq. (7.6)], the decrease in the factor $K(\varphi)$ overbalances the increase in $\Delta S_m$, so that $A_m$ continues to diminish. Thus, the amplitudes of the oscillations produced at point $P$ by Fresnel zones form a monotonously diminishing sequence:

$$A_1 > A_2 > A_3 > \cdots > A_{m-1} > A_m > A_{m+1} > \cdots$$

The phases of the oscillations produced by adjacent zones differ by $\pi$. Therefore, the amplitude $A$ of the resultant oscillation at point $P$ can be represented in the form

$$A = A_1 - A_2 + A_3 - A_4 + \cdots \tag{7.9}$$

This expression includes all the amplitudes from odd zones with one sign and from even zones with the opposite one .

Let us write Eq. (7.9) in the form

$$A = \frac{A_1}{2} + (\frac{A_1}{2} - A_2 + \frac{A_3}{2}) + (\frac{A_3}{2} - A_4 + \frac{A_5}{2}) + \cdots \quad (7.10)$$

Owing to the monotonous diminishing of $A_m$, we can approximately assume that

$$A_m = \frac{A_{m-1} + A_{m+1}}{2}$$

The expressions in parentheses will therefore vanish, and Eq (2. 10) will be simplified as follows:

$$A = \frac{A_1}{2} \quad (7.11)$$

According to Eq. (7.11), the amplitude produced at a point $P$ by an entire spherical wave surface equals half the amplitude produced by the central zone alone. If we put in the path of a wave an opaque screen having an aperture that leaves only the central Fresnel zone open, the amplitude at point $P$ will equal $A_1$, i.e. it will be double he amplitude given by Eq. (7.11). Accordingly, the intensity of the light at point $P$ will in this case be four times greater than when there are no barriers between points $S$ and $P$.

Now let us solve the problem on the propagation of light from source $S$ to point $P$ by the method of graphical addition of amplitudes. We shall divide the wave surface into annular zones similar to Fesnel zones, but much smaller in width (the path difference from the edges of a zone to point $P$ is a small fraction of $\lambda$ the same for all zones). We shall depict the oscillation produced at point $P$ by each of the zones in the form of a vector whose length equals the amplitude of the oscillation, while the angle made by the vector with the direction taken as the beginning of measurement gives the initial phase of the oscillation. The amplitude of the oscillations produced by such zones at point $P$ slowly diminishes from zone to zone. Each following oscillation lags behind the preceding one in phase by the same magnitude. Hence the vector diagram obtained when the oscillations produced by the separate zones are added has the form shown in Fig 7.6.

**Fig.7.6**

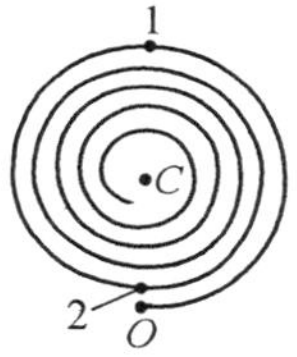


**Fig.7.7**

If the amplitudes produced by the individual zones were the same, the tail of the last of the vectors shown in Fig.7.6 would coincide with the tip of the first vector. Actually, the value of the amplitude diminishes, although very slightly. Hence, the vectors form a broken spiral-shaped line instead of a closed figure.

In the limit when the widths of the annular zones tend to zero (their number will grow unlimitedly), the vector diagram has the form of a spiral winding toward point $C$(Fig.7.7) The phases of the oscillations at points $O$ and 1 differ by $\pi$ (the infinitely small vectors forming the spiral have opposite directions at these points). Consequently, part $O$-1 of the spiral corresponds to the first Fresnel zone. The vector drawn from point $O$ to point 1 (Fig.7.8a) depicts the oscillation produced at point $P$ by this zone. Similarly, the vector drawn from point 1 to point 2 (Fig. 7.8b) depicts the oscillation produced by the second Fresnel zone. The oscillations from the first and second zones are in counterphase; accordingly, vectors 01 and 12 have opposite directions.

The oscillation produced at point $P$ by the entire wave surface is depicted by vector $OC$ (Fig.7.8c). Inspection of the figure shows that the amplitude in this case equals half the amplitude produced bu the first

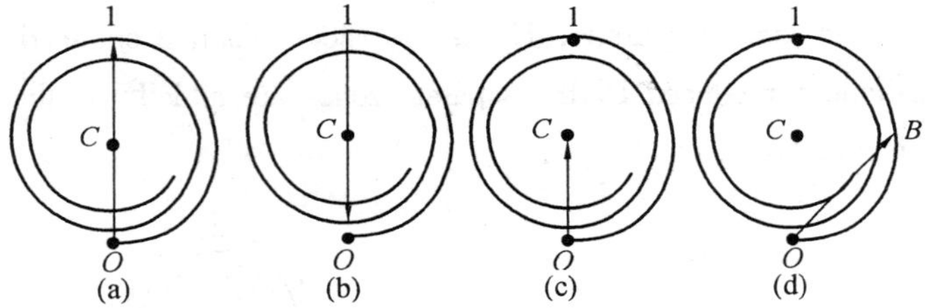


**Fig.7.8**

zone. We have obtained this result algebraically earlier. We shall note that the oscillation produced by the inner half of the first Fresnel zone is depicted by vector *OB* (Fig 7.8d). Thus, the action of the inner half of the first Fresnel zone is not equivalent to half the action of the first zone. Vector *OB* is $\sqrt{2}$ times greater than vector *OC*. Consequently, the intensity of the light produced by the inner half of the first Fresnel zone is double the intensity produced by the entire wave surface.

The oscillations from the even and odd Fresnel zones are in counterphase and, therefore, mutually weaken one another. If we would place in the path of the light wave a plate that would cover all the even or odd zones, the intensity of the light at point *P* would sharply grow. Such a plate, known as a zone one, functions like a converging lens. Figure 7.9 shows a plate covering the even zones. A still greater effect can be achieved by changing the phase of the even (or odd) zone oscillations by $\pi$ instead of covering these zones. This can be done with the aid of a transparent plate whose thickness at the places corresponding to the even or odd zones diners by a properly selected value. Such a plate is called a phase zone plate. In comparison with the amplitude zone plate covering zones, a phase plate produces an additional two-fold increase in the amplitude, and a four-fold increase in the light intensity.

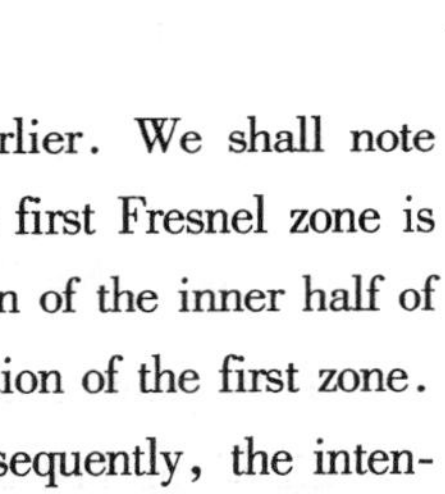

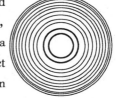

**Fig.7.9**

# The Further Study:

## Holography

Holography (i. e. "complete recording", from the Greek "holos" meaning "the whole" and "grapho" "write") is a special way of recording the structure of the light wave reflected by an object on a photographic plate. When this plate (a hologram) is illuminated with a beam of light, the wave recorded on it is reconstructed in practically its original form, so that when the eye perceives the reconstructed wave, the visual sensation is virtually the same as it would be if the object itself were observed.

Holography was invented in 1947 by the British physicist Dennis. Gabor. The complete embodiment of Gabor's idea became possible, however, only after the appearance in 1960 of light sources having a high degree of coherence-lasers, Gabor's initial arrangement was improved by the American physicists Emmet Leith and Juris Upatnieks, who obtained the first laser holograms in 1963. The Soviet scientist Yuri Denisyuk in 1962 proposed an original method of recording holograms on a thick-layer emulsion. This method, unlike holograms on a thin-layer emulsion, produces a colored image of the object.

We shall limit ourselves to an elementary consideration of the method of recording holograms on a thin-layer emulsion. Figure 7.10(a) contains a schematic view of an arrangement for recording holograms, and Fig. 7. 10(b) a schematic view of reconstruction of the image. The light beam emitted by laser, expanded by a system of lenses, is split into two parts. One part is reflected by the mirror to the photographic plate forming the so – called reference wave 1. The second part reaches the plate after being reflected from the object; it forms object beam 2. Both beams must be coherent. This requirement is satisfied because laser radiation has a high degree of spatial coherence (the light oscillations are coherent over the

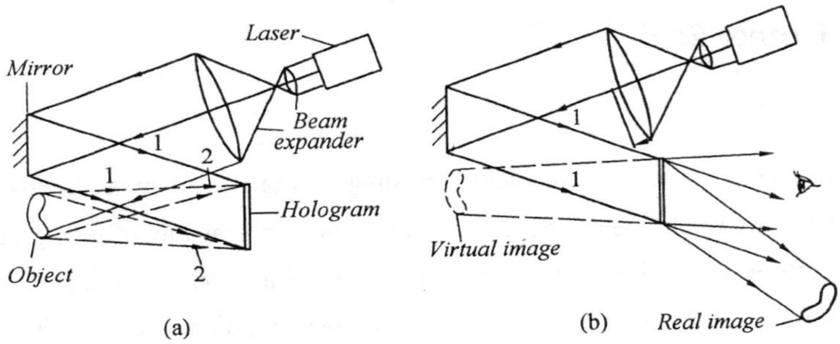


**Fig.7.10**

entire cross section of a laser beam). The reference and object beams superpose and form an interference pattern that is recorded by the photographic plate. A plate exposed in this way and developed is a hologram. Two beams of light participate in forming the hologram. In this connection, the arrangement described above is called two-beam or split-beam holography.

To reconstruct the image, the developed photographic plate is put in the same place where it was in recording the hologram, and is illuminated with the reference beam of light (the part of the laser beam that illuminated the object in recording the hologram is now stopped). The reference beam diffracts on the hologram, and as a result a wave is produced having exactly the same structure as the one reflected by the object. This wave produces a virtual image of the object that is seen by the observer. In addition to the wave forming the virtual image, another wave is produced that gives a real image of the object. This image is pseudoscopic; this means that it has a relief which is the opposite of the relief of the object—the convex spots are replaced by concave ones and vice versa.

Let us consider the nature of a hologram and the process of image reconstruction. Assume that two coherent parallel beams of light rays fall on the photographic plate, with the angle $\varphi$ between the beams (Fig.7.11). Beam 1 is the reference one. and beam 2, the object one (the object in the given case is an infinitely remote point). We shall assume for simplicity that beam 1 is normal to the plate. All the results obtained below also hold when the reference beam falls at an angle, but the formulas will be more cumbersome.

Owing to the interference of the reference and object beams, a system of alternating straight maxima and minima of the intensity is formed on the plate. Let points $A$ and $B$ correspond to the middles of adjacent interference maxima. Hence, the path difference $\Delta'$ equals $\lambda$. Examination of Fig. 7.11 shows that $\Delta' = d\sin\varphi$; hence,

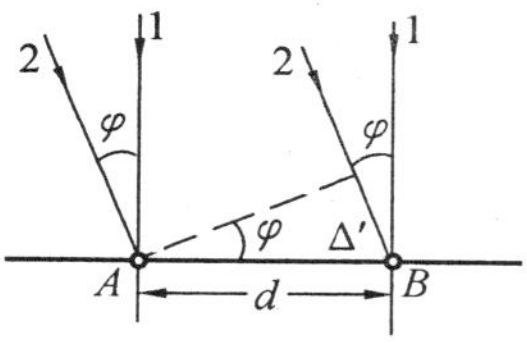


**Fig. 7.11**

$$d \sin \varphi = \lambda \tag{7.12}$$

Having recorded the interference pattern on the plate (by exposure and developing), we direct reference beam 1 at it. For this beam, the plate plays the part of a diffraction grating whose period $d$ is determined **by Eq. (7.12). A feature of** this grating is the circumstance that its transmittance changes in a direction perpendicular to the "lines" according to a cosine law. The result of this feature is that the intensity of all the diffraction maxima of orders higher than the first one virtually equals zero.

When the plate is illuminated with the reference beam (Fig. 7.12), a diffraction pattern appears whose maxima form the angles $\varphi$ with a normal to the plate. These angles are determined by the condition

$$d \sin \varphi = m\lambda \quad (m = 0, \pm 1) \tag{7.13}$$

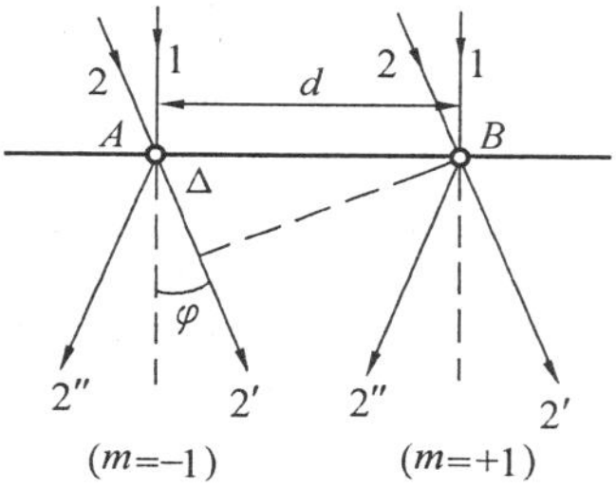


**Fig. 7.12**

The maximum corresponding to $m = 0$ is on the continuation of the reference beam. The maximum corresponding to $m = +1$ has the same direction as object beam 2 did during the exposure [compare Eqs. (7.12) and (7.13)]. In addition, a maximum corresponding to $m = -1$ appears.

It can be shown that the result we have obtained also holds when object beam 2 consists of diverging rays instead of parallel ones. The maximum corresponding to $m = +1$ has the nature of diverging beam of rays 2′ (it produces a virtual image of the point from which rays 2 emerged during the exposure); the maximum corresponding to $m = -1$, on the other hand, has the nature of a converging beam of rays 2″ (it forms a real image of the point which rays 2 emerged from during the exposure).

In recording the hologram, the plat is iuuminated by reference beam 1 and numerous diverging beams 2 reflected by different points of the object. An intricate interference pattern is formed on the plate as a result of superposition of the patterns produced by each of the beams 2 separately. When the hologram is illuminated with reference beam 1, all beams 2 are reconstructed, i.e. the complete light wave reflected by the object ($m = +1$ corresponds to it). Two other waves appear in addition to it (corresponding to $m = 0$ and $m = -1$). But these waves propagate in other directions and do not hinder the perception of the wave producing a virtual image of the object (see Fig.7.10).

The image of an object produced by a hologram is three-dimensional. It can be viewed from different positions. If in recording a hologram close objects concealed more remote ones, then by moving to a side we can look behind the closer object (more exactly, behind its image) and see the objects that had been concealed previously. The explanation is that when moving to a side we see the image reconstructed from the peripheral part of the hologram onto which the rays reflected from the concealed objects also fell during the exposure. When looking at the images of close and far objects, we have to accommodate our eyes as when looking at the objects themselves.

If a hologram is broken into several pieces, then each of them when illuminated will produce the same picture as the original hologram. But the smaller the part of the hologram used to reconstruct the image, the lower is its sharpness. This is easy to understand by taking into account that when the number of lines of a diffraction grating is reduced, its resolving power diminishes.

The possible applications of holography are very diverse. A far from complete list of them includes holographic motion pictures and television, holographic microscopes, and control of the quality of processing articles. The statement can be encountered in publications on the subject that holography can be compared regards its consequences with the setting up of radio communication.

### Summary of Key Terms

**Diffraction** The bending of light around an obstacle or through a narrow slit in such a way that fringes of light and dark or colored bands are produced.

**Polarization** The alignment of the electric vectors that make up electromagnetic radiation. Such waves of aligned vibrations are said to be polarized.

**Hologram** A two-dimensional microscopic diffraction pattern that shoes three-dimensional optical images.

## Reading Materials:

### Why the Sky Is Blue

If a beam of a particular frequency of sound is directed to a tuning fork of similar frequency, the tuning fork will be set into vibration and will effectively redirect the beam in many directions. The tuning fork will be set into vibration and will effectively redirect the beam in many directions. The tuning fork scatters the sound. A similar process occurs with the scattering of light from atoms and particles that are far apart from one another, as in the atmosphere.

We know that atoms behave like tiny optical tuning forks and re-emit light waves that shine on them. Very tiny particles do the same. The tinier the particle, the higher the frequency of light it will scatter. This is similar to the way small bells ring with higher notes than larger bells. The nitrogen and oxygen molecules and the tiny particles that make up the atmosphere are like tiny bells that "ring" with high frequencies when energized by sunlight. Like sound from the bells, the re-emitted light is sent in all directions. It is scattered.

Most of the ultraviolet light from the sun is absorbed by a thin protective layer of ozone gas in the upper atmosphere. The remaining ultraviolet sunlight that passes through the atmosphere is scattered by atmospheric particles and molecules. Of the visible frequencies, violet is scattered the most, followed by blue, green, yellow, orange, and red, in that order. Red is scattered only a tenth as much as violet. Although violet light is scattered more than blue, our eyes are not very sensitive to violet light. The lesser amount of blue predominates in our vision, so we see a blue sky!

The blue of the sky varies in different places under different conditions. A principal factor is the water-vapor content of the atmosphere. On clear dry days the sky is a much deeper blue than on clear days with high humidity. Places where the upper air is exceptionally dry, such as Italy and Greece, have beautifully blue skies that have inspired painters for centuries. Where there are a lot of particles of dust and other particles larger than oxygen and nitrogen molecules, the particles have been washed away, the sky becomes a deeper blue.

The grayish haze in the skies of large cities is the result of particles emitted by internal combustion engines (cars, trucks, and industrial plants). Even when idling, a typical automobile engine emits more than 100 billion particles per second. Most are invisible and provide a framework to which other particles adhere. These are the primary scatterers of lower frequency light. For the larger of these particles, absorption rather than scattering takes place and a brownish haze is produced. Yuk!

## Why Clouds Are White

Clusters of water molecules in a variety of sizes make up clouds. The different size clusters result in a variety of scattered frequencies: the tiniest, blue; slightly larger clusters, say, green; and still larger clusters, red. The overall result is a white cloud. Electrons close to one another in a cluster vibrate together and in step, which results in a greater intensity of scattered light than from the same number of electrons vibrating separately. Hence, clouds are bright!

Absorption occurs for larger droplets, and the scattered intensity is less. The clouds are darker. Further increase in the size of the drops causes them to fall to earth, and we have rain.

The next time you find yourself admiring a crisp blue sky, or delighting in the shapes of bright clouds, or watching a beautiful sunset, think about all those ultra-tiny optical tuning forks vibrating away—you'll appreciate these everyday wonders of nature even more!

# PART 4
# ELECTRICITY and MAGNETISM

# 8

## Electric Field in a Vacuum

### 8.1 ELECTRIC CHARGE

All bodies in nature are capable of becoming electrified, i.e. acquiring an electric charge. The presence of such a charge manifests itself in that a charged body interacts with other charged bodies. Two kinds of electric charges exist. They are conventionally called positive and negative. Like charges repel each other, and unlike charges attract each other.

An electric charge is an integral part of certain elementary particles. The charge of all elementary particles (if it is not absent) is identical in magnitude. It can be called an elementary charge. We shall use the symbol $e$ to denote a positive elementary charge.

The elementary particles include, in particular, the electron (carrying the negative charge $-e$) the proton (carrying the positive charge

$+e$), and the neutron (carrying no charge). These particles are the bricks which the atoms and molecules of any substance are built of, therefore all bodies contain electric charges. The particles carrying charges of different signs are usually present in a body in equal numbers and are distributed over it with the same density. The algebraic sum of the charges in any elementary volume of the body equals zero in this case, and each such volume (as well as the body as a whole) will be neutral. If in some way or other we create a surplus of particles of one sign in a body (and, correspondingly, a shortage of particles of the opposite sign), the body will be charged. It is also possible, without changing the total number of positive and negative particles, to cause them to be redistributed in a body so that one part of it has a surplus of charges of one sign and the other part a surplus of charges of the opposite sign. This can be done by bringing a charged body close to an uncharged metal one.

Since a charge $q$ is formed by a plurality of elementary charges, it is an integral multiple of $e$:

$$q = \pm N\mathrm{e} \tag{8.1}$$

An elementary charge is so small, however, that macroscopic charges may be considered to have continuously changing magnitudes.

If a physical quantity can take on only definite discrete value, it is said to be quantized. The fact expressed by Eq. (8.1) signifies that an electric charge is quantized.

The magnitude of a charge measured in different inertial reference frames will be found to be the same. Hence, an electric charge is relativistically invariant. It thus follows that the magnitude of a charge does not depend on whether the charge is moving or at rest.

Electric charges can vanish and appear again. Two elementary charges of opposite signs always appear or vanish simultaneously, however. For example, an electron and a positron (a positive electron) meeting each other annihilate, i.e. transform into neutral gamma-photons. This is attended by vanishing of the charge $-e$ and $+e$. In the course of the process called the birth of a pair, a gamma-photon getting into the field of an atomic nucleus transforms into a pair of particles an electron and a positron. This process causes the charges $-e$ and $+e$ to appear.

Thus, the total charge of an electrically isolated system cannot change. This statement forms the law of electric charge conservation.

We must note that the law of electric charge conservation is associated very closely with the relativistic invariance of a charge. Indeed, if the magnitude of a charge depended on its velocity, then by bringing charges of one sign into motion we would change the total charge of the relevant isolated system.

## 8.2 COULOMB'S LAW

The law obeyed by the force of interaction of point charges was established experimentally in 1785 by the French physicist Charles. A. de Coulomb (1736 ~ 1806). A point charge is defined as a charged body whose dimensions may be disregarded in comparison with the distances from this body to other bodies carrying an electric charge.

Using a torsion balance (Fig. 8.1) similar to that employed by H. Cavendish to determine the gravitational constant , Coulomb measured the force of interaction of two charged spheres depending on the magnitude of the charges on them and on the distance between them. He proceeded from the fact that when a charged metal sphere was touched by an identical uncharged sphere, the charge would be distributed equally between the two spheres.

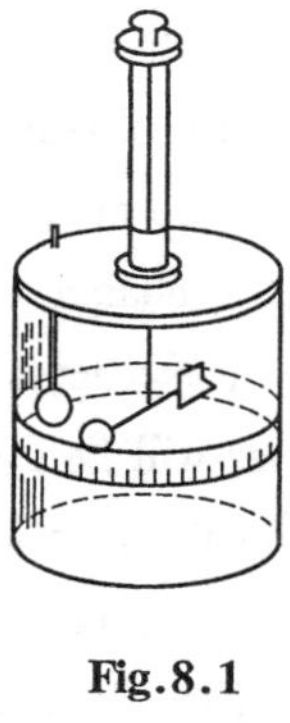

**Fig.8.1**

As a result of his experiments, Coulomb arrived at the conclusion that the force of interaction between two stationary point charges is proportional to the magnitude of each of them and inversely proportional to the square of the distance between them. The direction of the force coincides with the straight line connecting the charges.

It must be noted that the direction of the force of interaction along the straight line connecting the point charges follows from considerations of symmetry. An empty space is assumed to be homogeneous and isotropic. Consequently, the only direction distinguished in the space by stationary point charges introduced into it is that from one charge to the other. Assume that the force $\boldsymbol{F}$ acting on the charge $q_1$(Fig. 8.2) makes the angle $\alpha$ with the direction from $q_1$ to $q_2$ and that $\alpha$ differs from 0 or $\pi$. But owing to axial symmetry, there are no grounds to set the force $\boldsymbol{F}$ aside from the multitude of forces of other directions making the same angle $\alpha$ with the axis $q_1$-$q_2$ (the directions of these forces form a cone with a cone angle of $2\alpha$). The difficulty appearing as a result of this vanishes when $\alpha$ equals 0 or $\pi$.

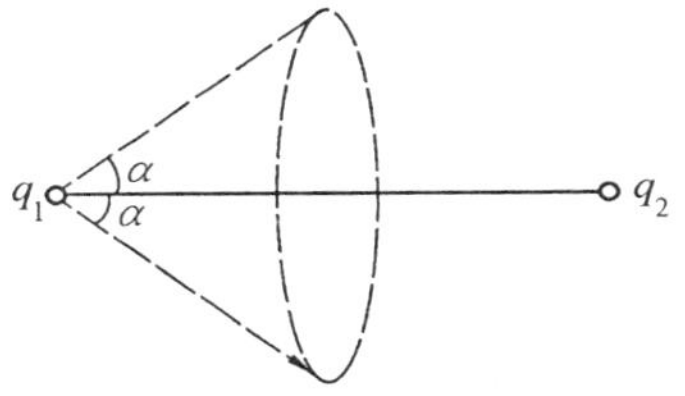


**Fig.8.2**

Coulomb's law can be expressed by the formula

$$\boldsymbol{F}_{12} = -k\frac{q_1 q_2}{r^2}\boldsymbol{e}_{12} \tag{8.2}$$

Here $k$ = proportionality constant assumed to be positive

$q_1$ and $q_2$ = magnitudes of the interacting charges

$r$ = distance between the charges

$\boldsymbol{e}_{12}$ = unit vector directed from the charge $q_1$ to $q_2$

$\boldsymbol{F}_{12}$ = force acting on the charge $q_1$(Fig.8.3; the figure corresponds to the case of like charges).

The force $\boldsymbol{F}_{12}$ differs from $\boldsymbol{F}_{21}$ in its sign:

$$\boldsymbol{F}_{21} = k\frac{q_1 q_2}{r^2}\boldsymbol{e}_{12} \tag{8.3}$$

The magnitude of the interaction force, which is the same for both charges, can be written in the form

$$F = k \frac{|q_1 q_2|}{r^2} \tag{8.4}$$

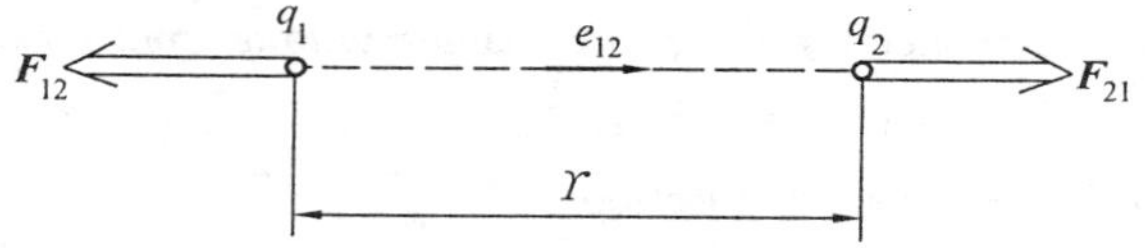


**Fig.8.3**

Experiments show that the force of interaction between two given charges does not change if other charges are placed near them. Assume that we have the charge $q_a$ and, in addition, $N$ other charges $q_1, q_2, \cdots, q_N$. It can be seen from the above that the resultant force $\boldsymbol{F}$ with which all the $N$ charge $q_i$ act on $q_a$ is

$$\boldsymbol{F} = \sum_{i=1}^{N} \boldsymbol{F}_{a,i} \tag{8.5}$$

where $\boldsymbol{F}_{a,i}$ is the force with which the charge $q_i$ acts on $q_a$ in the absence of the other $N-1$ charges.

The fact expressed by Eq. (8.5) permits us to calculate the force of interaction between charges concentrated on bodies of finite dimensions, knowing the law of interaction between point charges. For this purpose, we must divide each charge into so small charger $dq$ that they can be considered as point ones, use Eq. (8.2) to calculate the force of interaction between the charges $dq$ taken in pairs, and then perform vector summation of these forces. Mathematically, this procedure coincides completely with the calculation of the force of gravitational attraction between bodies of finite dimensions.

All experimental facts available lead to the conclusion that Coulo – mb's law holds for distances from $10^{-15}$ m to at least several kilometers. There arc grounds to presume that for distances smaller than $10^{-16}$ m the law stops being correct. For very great distances, there are no experimental confirmations of Coulomb's law. But there are also no reasons to expect that this law stops being obeyed with very great distances between charges.

## 8.3 SYSTEMS OF UNITS

We can make the proportionality constant in Eq. (8.2) equal unity by properly choosing the unit of charge (the units for $F$ and $r$ were established in mechanics). The relevant unit of charge (when $F$ and $r$ are measured in cgs units) is called the absolute electrostatic unit of charge ($\text{cgse}_q$). It is the magnitude of a charge that interacts with a force of 1 dyne in a vacuum with an equal charge at a distance of 1 cm from it.

Careful measurements showed that an elementary charge is

$$e = 4.80 \times 10^{-10}\ \text{cgse}_q \tag{8.6}$$

Adopting the units of length, mass, time, and charge as the basic ones, we can construct a system of units of electrical and magnetic quantities. The system based on the centimeter, gramme, second; and the $\text{cgse}_q$ unit is called the absolute electrostatic system of units (the cgse system). It is founded on Coulomb's law, i.e. the law of interaction between charges at rest. On a later page, we shall become acquainted with the absolute electromagnetic system of units (the cgsm system) based on the law of interaction between-conductors carrying an electric current. The Gaussian system in which the units of electrical quantities coincide with those of the cgse system, and of magnetic quantities with those of the cgsm system, is also an absolute system.

Equation (8.4) in the cgse system becomes

$$F = \frac{|q_1 q_2|}{r^2} \tag{8.7}$$

This equation is correct if the charge are in a vacuum. It has to be determined more accurately for charges in a medium.

USSR State Standard GOST 9867 - 61, which came into force on

January 1. 1963. prescribes the preferable use of the International System of Units (SI). The basic units of this system are the meter, kilogramme, second. ampere, kelvin. candela, and mole. The SI unit of force is the newton ($N$) equal to $10^5$ dynes.

In establishing the units of electrical and magnetic quantities, the SI system, like the cgsm one, proceeds from the law of interaction of current – carrying conductors instead of charges. Consequently, the proportionality constant in the equation of Coulomb's law is a quantity with a dimension and differing from unity.

The SI unit of charge is the coulomb (C). It has been found experimentally that

$$1\text{C} = 2.998 \times 10^9 \approx 3 \times 10^9 \text{ cgse}_q \tag{8.8}$$

To form an idea of the magnitude of a charge of 1 C, let us calculate the force with which two point charges of 1 C each would interact with each other if they were 1 m apart. By Eq.(8.7)

$$F = \frac{3 \times 10^9 \times 3 \times 10^9}{100^2} \text{ cgse}_F = 9 \times 10^{14} \text{ dyn} = 9 \times 10^9 \text{ N} \approx 10^9 \text{ kgf} \tag{8.9}$$

An elementary charge expressed in coulombs is

$$e = 1.60 \times 10^{-19} \text{ C} \tag{8.10}$$

## 8.4 RATIONALIZED FORM OF WRITING FORMULAS

Many formulas of electrodynamics when written in the cgs systems (in particular, in the Gaussian one) include as factors $4\pi$ and the so-called electromagnetic constant c equal to the speed of light in a vacuum. To eliminate these factors in the formulas that are most important in practice, the proportionality constant in Coulomb's law is taken equal to $1/4\pi\varepsilon_0$ The equation of the law for charges in a vacuum will thus become

$$F = \frac{1}{4\pi\varepsilon_0} \frac{|q_1 q_2|}{r^2} \tag{8.11}$$

The other formulas change accordingly. This modified way of writing formulas is called rationalize. Systems of units constructed with the use of rationalized formulas are also called rationalized. They include the SI system.

The quantity $\varepsilon_0$ is called the electric constant. It has the dimension of capacitance divided by length. It is accordingly expressed in units called the farad per meter. To find the numerical value of $\varepsilon_0$, we shall introduce the values of the quantities corresponding to the case of two charges 1 C each and 1 m apart into Eq. (8.11). By Eq. (1.9), the force of interaction in this case is $9 \times 10^9$ N. Using this value of the force, and also $q_1 = q_2 = 1\text{C}$ and $r = 1$ m in Eq. (8.11), we get

$$9 \times 10^9 = \frac{1}{4\pi\varepsilon_0}\frac{1 \times 1}{1^2}$$

whence

$$\varepsilon_0 = \frac{1}{4\pi \times 9 \times 10^9} = 0.885 \times 10^{-11}\ F/m \qquad (8.12)$$

The Gaussian system of units was widely used and is continuing to be used in physical publications. We therefore consider it essential to acquaint our reader with both the SI and the Gaussian system. We shall set out the material in the SI units showing at the same time how the formulas look Gaussian system.

## 8.5 ELECTRIC FIELD. FIELD STRENGTH

Charge at rest interact through an electric field. A charge alters the properties of the space surrounding it-it sets up an electric field in it. This field manifests itself in that an electric charge placed at a point of it experiences the action of a force. Hence, to see whether there is an electric field at a given place, we must place a charged body (in the following we shall say simply a charge for brevity) at it and determine whether or not it experiences the action of an electric force. We can evidently assess the "strength" of the field according to the magnitude of the force exerted on the given charge.

Thus, to detect and study an electric field, we must use a "test" charge. For the force acting on our test charge to characterize the field "at the given point", the test charge must be a point one. Otherwise, the force acting on the charge will characterize the properties of the field averaged over the volume occupied by the body that carries the test charge.

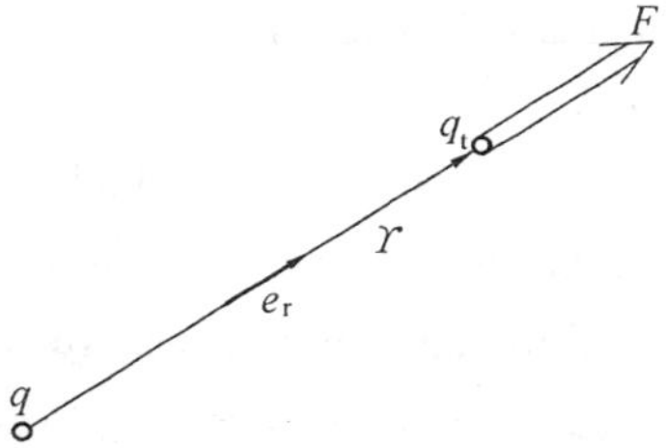

**Fig.8.4**

Let us study the field set up by the stationary point charge $q$ with the aid of the point test charge $q_t$. We place the test charge at a point whose position relative to the charge $q$ is determined by the position vector $\boldsymbol{r}$ (Fig.1.4). We see that the test charge experiences the force

$$\boldsymbol{F} = q_t\left(\frac{1}{4\pi\varepsilon_0}\frac{q}{r^2}\boldsymbol{e}_r\right) \tag{8.13}$$

[see Eqs. (8.3) and (8.11)]. Here $\boldsymbol{e}_r$, is the unit vector of the position vector $\boldsymbol{r}$.

A glance at Eq. (8.13) shows that the force acting on our test charge depends not only on the quantities determining the field (on $q$ and $\boldsymbol{r}$), but also on the magnitude of the test charge $q_t$ If we take different test charges $q'_t$, $q''_t$, etc., then the forces $\boldsymbol{F}'$, $\boldsymbol{F}''$, etc. which they experience at the given point of the field will be different. We can see from Eq.(8.13), however, that the ratio $\boldsymbol{F}/q_t$ for all the test charges will be the same and depend only on the values of $q$ and $\boldsymbol{r}$ determining the field at the given point. It is therefore natural to adopt this ratio as the quantity characterizing an electric field:

$$\boldsymbol{E} = \frac{\boldsymbol{F}}{q_t} \tag{8.14}$$

This vector quantity is called the electric field strength (or intensity) at a given point (i.e. at the point where the test charge $q_t$ experiences the action of the force $\boldsymbol{F}$).

According to Eq. (8.14), the electric field strength numerically equals the force acting on a unit point charge at the given point of the field. The direction of the vector $\boldsymbol{E}$ coincides with that of the force acting on a positive charge.

It must be noted that Eq. (8.14) also holds when the test charge is negative ($q_t < 0$). In this case, the vectors $\boldsymbol{E}$ and $\boldsymbol{F}$ have opposite directions.

We have arrived at the concept of electric field strength when studying the field of a stationary point charge. Definition (8.14), however, also covers the case of a field set up by any collection of stationary charges. But here the following clarification is needed. The arrangement of the charges setting up the field being studied may change under the action of the test charge. This will happen, for example, when the charges producing the field are on a conductor and can freely move within its limits. Therefore, to avoid appreciable alterations in the field being studied, a sufficiently small test charge must be taken.

It follows from Eqs. (8.14) and (8.13) that the field strength of a point charge varies directly with the magnitude of the charge $q$ and inversely with the square of the distance $r$ from the charge to the given point of the field:

$$\boldsymbol{E} = \frac{1}{4\pi\varepsilon_0}\frac{q}{r^2}\boldsymbol{e}_r \qquad (8.15)$$

The vector $\boldsymbol{E}$ is directed along the radial straight line passing through the charge and the given point of the field, from the charge if the latter is positive and toward the charge if it is negative.

In the Gaussian system, the equation for the field strength of a point charge in a vacuum has the form

$$E = \frac{q}{r^2}\mathbf{e}_r \tag{8.16}$$

The unit of electric field strength is the strength at a point where unit force (1 N in the SI and 1 dyn in the Gaussian system) acts on unit charge (1 C in the SI and 1 $cgse_q$ in the Gaussian system). This unit has no special name in the Gaussian system. The SI unit of electric field strength is called the volt per meter (V/m) [see Eq. (8.44)].

According to Eq. (8.15), a charge of 1 C produces the following field strength in a vacuum at a distance of 1 m from this charge:

$$E = \frac{1}{4\pi(1/4\pi \times 9 \times 10^9)}\frac{1}{1^2} = 9 \times 10^9 \text{ V/m}$$

This strength in the Gaussian system is

$$E = \frac{q}{r^2} = \frac{3 \times 10^9}{100^2} = 3 \times 10^5 \text{ cgse}_E$$

Comparing these two results, we find that

$$1 \text{ cgse}_E = 3 \times 10^4 \text{ V/m} \tag{8.17}$$

According to Eq. (8.14), the force exerted on a test charge is

$$\boldsymbol{F} = q_t\boldsymbol{E}$$

It is obvious that any point charge $q$ at a point of a field with the strength $\boldsymbol{E}$ will experience the force

$$\boldsymbol{F} = q\boldsymbol{E} \tag{8.18}$$

If the charge $q$ is positive, the direction of the force coincides with that of the vector $\boldsymbol{E}$. If $q$ is negative, the vectors $\boldsymbol{F}$ and $\boldsymbol{E}$ are directed oppositely.

We mentioned in Sec. 8.2 that the force with which a system of charges acts on a charge not belonging to the system equals the vector sum of the forces which each of the charges of the system exerts separately on the given charge [see Eq. (8.5)]. Hence it follows that the field strength of a system of charges equals the vector sum of the field strengths that would be produced by each of the charges of the system separately:

$$\boldsymbol{E} = \sum \boldsymbol{E}_i \tag{8.19}$$

This statement is called the principle of electric field superposition.

The superposition principle allows us to calculate the field strength of any system of charges. By dividing extended charges into sufficiently small fractions $\mathrm{d}q$, we can reduce any system of charges to a collection of point charges. We calculate the contribution of each of such charges to the resultant field by Eq. (8.15).

An electric field can be described by indicating the magnitude and direction of the vector $\boldsymbol{E}$ for each of its points. The combination of these vectors forms the field of the electric field strength vector. The velocity vector field can be represented very illustratively with the aid of flow lines. Similarly, an electric field can be described with the aid of strength lines, which we shall call for short $\boldsymbol{E}$ lines or field lines. These lines are drawn so that a tangent to them at every point coincides with the direction of the vector $\boldsymbol{E}$. The density of the lines is selected so that their number passing through a unit area at right angles to the lines equals the numerical value of the vector $\boldsymbol{E}$. Hence, the pattern of field lines permits us to assess the direction and magnitude of the vector $\boldsymbol{E}$ at various points of space (Fig. 8.5).

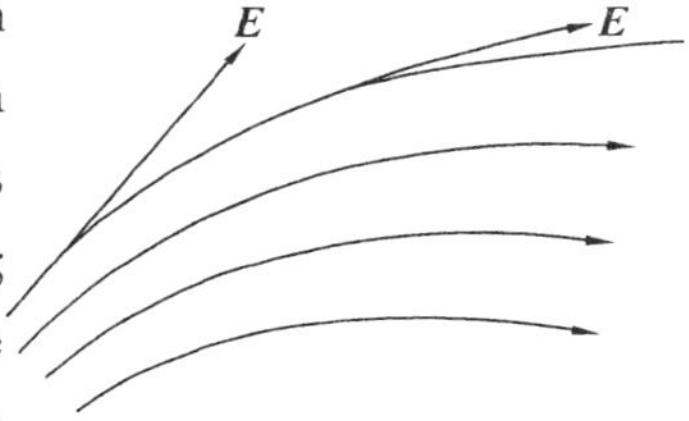


**Fig. 8.5**

The $\boldsymbol{E}$ lines of a point charge field are a collection of radial straight lines directed away from the charge if it is positive and toward it if it is negative (Fig. 8.6). One end of each line is at the charge, and the other extends to infinity. Indeed, the total number of lines intersecting a spherical surface of arbitrary radius $r$ will equal the product of the density of the lines and the surface area of the sphere $4\pi r^2$ We have assumed that the density of the lines numerically equals $E = (1/4\pi\varepsilon_0)(q/r^2)$.

Hence, the number of lines is $(1/4\pi\varepsilon_0)(q/r^2)4\pi r^2 = q/\varepsilon_0$. This result

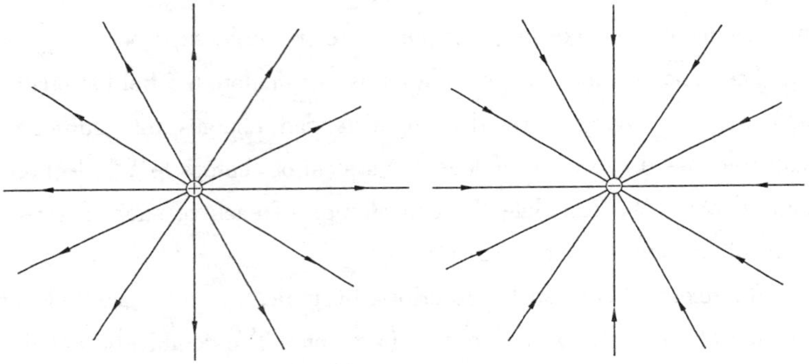

**Fig.8.6**

signifies that the number of lines at any distance from a charge will be the same. It thus follows that the lines do not begin and do not terminate anywhere except for the charge. Beginning at the charge, they extend to infinity (the charge is positive), or arriving from infinity, they terminate at the charge (the latter is negative). This property of the $\boldsymbol{E}$ lines is common for all electrostatic fields, i.e. fields set up by any system of stationary charges: the field lines can begin or terminate only at charges or extend to infinity.

## 8.6 POTENTIAL

Let us consider the field produced by a stationary point charge $q$. At any point of this field, the point charge $q'$ experiences the force

$$\boldsymbol{F} = \frac{1}{4\pi\varepsilon_0}\frac{qq'}{r^2}\boldsymbol{e}_r = F(r)\boldsymbol{e}_r \tag{8.20}$$

Here $F(r)$ is the magnitude of the force $\boldsymbol{F}$, and $\boldsymbol{e}_r$ the unit vector of the position vector $\boldsymbol{r}$ determining the position of the charge $q'$ relative to the charge $q$.

The force (8.20) is a central one. A central field of forces is conservative. Consequently, the work done by the forces of the field on the charge $q'$ when it is moved from one point to another does not depend on the path. This work is

$$\boldsymbol{A}_{12} = \int_1^2 \boldsymbol{F}(r)\boldsymbol{e}_r \mathrm{d}\boldsymbol{l} \quad (8.21)$$

where $\mathrm{d}\boldsymbol{l}$ is the elementary displacement of the charge $q'$. Inspection of Fig. 8.7 shows that the scalar product $\boldsymbol{e}_r \mathrm{d}\boldsymbol{l}$ equals the increment of the magnitude of the position vector $\boldsymbol{r}$, i.e. $\mathrm{d}r$. Equation (8.21) can therefore be written in the form

$$A_{12} = \int_1^2 F(r)\mathrm{d}r$$

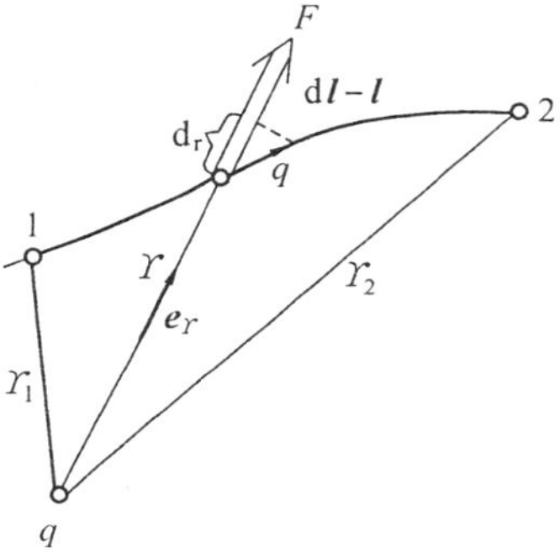


**Fig. 8.7**

Introduction of the expression for $F(r)$ yields

$$A_{12} = \frac{qq'}{4\pi\varepsilon_0}\int_{r_1}^{r_2}\frac{\mathrm{d}r}{r^2} = \frac{1}{4\pi\varepsilon_0}\left(\frac{qq'}{r_1} - \frac{qq'}{r_2}\right) \quad (8.22)$$

The work of the forces of a conservative field can be represented as a decrement of the potential energy:

$$A_{12} = W_{p,1} - W_{p,2} \quad (8.23)$$

A comparison of Eqs. (8.22) and (8.23) leads to the following expression for the potential energy of the charge $q'$ in the field of the charge $q$:

$$W_p = \frac{1}{4\pi\varepsilon_0}\frac{qq'}{r} + \text{const}$$

The value of the constant in the expression for the potential energy is usually chosen so that when the charge moves away to infinity (i.e. when $r \to \infty$), the potential energy vanishes. When this condition is observed, we get

$$W_p = \frac{1}{4\pi\varepsilon_0}\frac{qq'}{r} \quad (8.24)$$

Let us use the charge $q'$ as a test charge for studying the field. By Eq. (8.24), the potential energy which the test charge has depends not only on its magnitude $q'$, but also on the quantities $q$ and $r$ determining the field. Thus, we can use this energy to describe the field just like we used the force acting on the test charge for this purpose.

Different test charges $q'$, $q''_t$. etc. will have different energies $W'_p$, $W''_p$, etc. at the same point of a field. But the ratio $W_p/q_t$ will be the same for all the charges [see Eq (8.24)]. The quantity

$$\varphi = \frac{W_p}{q_t} \tag{8.25}$$

is called the field potential at a given point and is used together with the field strength $E$ to describe electric fields.

It can be seen from Eq. (8.25) that the potential numerically equals the potential energy which a unit positive charge would have at the given point of the field. Substituting for the potential energy in Eq. (8.25) its value from (8.24), we get the following expression for the potential of a point charge:

$$\varphi = \frac{1}{4\pi\varepsilon_0} \cdot \frac{q}{r} \tag{8.26}$$

In the Gaussian system, the potential of the field of a point charge in a vacuum is determined by the formula

$$\varphi = \frac{q}{r} \tag{8.27}$$

Let us consider the field produced by a system of $N$ point charges $q_1, q_2 \cdots, q_N$. Let $r_1, r_2, \cdots r_N$ be the distances from each of the charges to the given point of the field. The work done by the forces of this field on the charge $q'$ will equal the algebraic sum of the work done by the forces set up by each of the charges separately:

$$A_{12} = \sum_{i=1}^{N} A_i$$

By Eq. (8.22), each work $A_i$ equals

$$A_i = \frac{1}{4\pi\varepsilon_0}\left(\frac{q_i q'}{r_{i,1}} - \frac{q_i q'}{r_{i,2}}\right)$$

where $r_{i,1}$ is the distance from the charge $q_i$ to the initial position of the charge $q'$ and $r_{i,2}$ is the distance from $q_i$ to the final position of the charge $q'$. Hence,

$$A_{12} = \frac{1}{4\pi\varepsilon_0}\sum_{i=1}^{N}\frac{q_i q'}{r_{i,1}} - \frac{1}{4\pi\varepsilon_0}\sum_{i=1}^{N}\frac{q_i q'}{r_{i,2}}$$

Comparing this equation with Eq. (8.23), we get the following expression for the potential energy of the charge $q'$ in the field of a system of charges:

$$W_p = \frac{1}{4\pi\varepsilon_0}\sum_{i=1}^{N}\frac{q_i q'}{r_i}$$

from which it can be seen that

$$\varphi = \frac{1}{4\pi\varepsilon_0}\sum_{i=1}^{N}\frac{q_i}{r_i} \qquad (8.28)$$

Comparing this formula with Eq. (8.26), we arrive at the conclusion that the potential of the field produced by a system of charges equals the algebraic sum of the potentials produced by each of the charges separately. Whereas the field strengths are added vectorially in the superposition of fields, the potentials are added algebraically. This is why it is usually much simpler to calculate the potentials than the electric field strengths.

Examination of Eq. (8.25) shows that the charge $q$ at a point of a field with the potential $\varphi$ has the potential energy

$$W_p = q\varphi \qquad (8.29)$$

Hence, the work of the field forces on the charge $q$ can be expressed through the potential difference:

$$A_{12} = W_{p,1} - W_{p,2} = q(\varphi_1 - \varphi_2) \qquad (8.30)$$

Thus, the work done on a charge by the forces of a field equals the product of the magnitude of the charge and the difference between the potentials at the initial and final points (i.e. the potential decrement).

If the charge $q$ is removed from a point having the potential $\varphi$ to infinity (where by convention the potential vanishes), then the work of the field forces will be

$$A_{\infty} = q\varphi \tag{8.31}$$

Hence, it follows that the potential numerically equals the work done by the forces of a field on a unit positive charge when the latter is removed from the given point to infinity. Work of the same magnitude must be done against the electric field forces to move a unit positive charge from infinity to the given point of a field.

Equation (8.31) can be used to establish the units of potential. The unit of potential is taken equal to the potential at a point of a field when work equal to unity is required to move unit positive charge from infinity to this point. The SI unit of potential called the volt (V) is taken equal to the potential at a point when work of 1 joule has to be done to move a charge of 1 coulomb from infinity to this point:

$$1\ \text{J} = 1\ \text{C} \times 1\ \text{V}$$

whence

$$1\ \text{V} = \frac{1\ \text{J}}{1\ \text{C}} \tag{8.32}$$

The absolute electrostatic unit of potential ($\text{cgse}_{\varphi}$) is taken equal to the potential at a point when work of 1 erg has to be done to move a charge of $+1\ \text{cgse}_q$ from infinity to this point. Expressing 1 J and 1 C in Eq. (8.32) through cgseq units, we shall find the relation between the volt and the cgse potential unit:

$$1\ \text{V} = \frac{1\ \text{J}}{1\ \text{C}} = \frac{10^7\ \text{erg}}{3 \times 10^9\ \text{cgse}_q} = \frac{1}{300}\ \text{cgse}_{\varphi} \tag{8.33}$$

Thus, $1\ \text{cgse}_{\varphi}$ equals 300 V.

A unit of energy and work called the electron-volt (eV) is frequently used in physics. An electron-volt is defined as the work done by the forces of a field on a charge equal to that of an electron (i.e. on the elementary charge $e$) when it passes through a potential difference of 1 V:

$$1\ \text{eV} = 1.60 \times 10^{-19}\ \text{C} \times 1\ \text{V} = 1.60 \times 10^{-19}\ \text{J} = 1.60 \times 10^{-12}\ \text{erg} \tag{8.34}$$

Multiple units of the electron-volt are also used:

1 keV (kiloelectron-volt) = $10^3$ eV

1 MeV (megaelectron-volt) = $10^6$ eV

1 GeV (gigaelectron-volt) = $10^9$ eV

## 8.7 INTERACTION ENERGY OF A SYSTEM OF CHARGES

Equation (8.24) can be considered as the mutual potential energy of the charges $q$ and $q'$. Using the symbols $q_1$ and $q_2$ for these charges, we get the following formula for their interaction energy:

$$W_p = \frac{1}{4\pi\varepsilon_0}\frac{q_1 q_2}{r_{12}} \tag{8.35}$$

The symbol $r_{12}$ stands for the distance between the charges.

Let us consider a system consisting of $N$ point charges $q_1, q_2 \cdots, q_N$. We showed that the energy of interaction of such a system equals the sum of the energies of interaction of the charges taken in pairs:

$$W_p = \frac{1}{2}\sum_{(i \neq k)} W_{p,ik}(r_{ik}) \tag{8.36}$$

According to Eq. (8.35)

$$W_{p,ik} = \frac{1}{4\pi\varepsilon_0}\frac{q_i q_k}{r_{ik}}$$

Using this equation in (8.36), we find that

$$W_p = \frac{1}{2}\sum_{i \neq k}\frac{1}{4\pi\varepsilon_0}\frac{q_i q_k}{r_{ik}} \tag{8.37}$$

In the Gaussian system, the factor $1/4\pi\varepsilon_0$ is absent in this equation.

In Eq. (8.37), summation is performed over the subscripts $i$ and $k$. Both subscripts pass independently through all the values from 1 to $N$. Addends for which the value of the subscript $i$ coincides with that of $k$ are not taken into consideration. Let us write Eq. (8.37) as follows:

$$W_p = \frac{1}{2}\sum_{i=1}^{N} q_i \sum_{k=1, k\neq i}^{N} \frac{1}{4\pi\varepsilon_0}\frac{q_k}{r_{ik}} \qquad (8.38)$$

The expression

$$\varphi_i = \frac{1}{4\pi\varepsilon_0}\sum_{k=1, k\neq i}^{N}\frac{q_k}{r_{ik}}$$

is the potential produced by all the charges except $q_i$ at the point where the charge $q_i$ is. With this in view, we get the following formula for the interaction energy:

$$W_p = \frac{1}{2}\sum_{i=1}^{N} q_i\varphi_i \qquad (8.39)$$

## 8.8 RELATION BETWEEN ELECTRIC FIELD STRENGTH AND POTENTIAL

An electric field can be described either with the aid of the vector quantity $\boldsymbol{E}$, or with the aid of the scalar quantity $\varphi$. There must evidently be a definite relation between these quantities. If we bear in mind that $\boldsymbol{E}$ is proportional to the force acting on a charge and $\varphi$ to the potential energy of the charge, it is easy to see that this relation must be similar to that between the potential energy and the force.

The force $\boldsymbol{F}$ is related to the potential energy by the expression

$$\boldsymbol{F} = -\nabla W_p \qquad (8.40)$$

For a charged particle in an electrostatic field, we have $\boldsymbol{F} = q\boldsymbol{E}$ and $W_p = q\varphi$. Introducing these values into Eq. (8.40), we find that

$$q\boldsymbol{E} = -\nabla(q\varphi)$$

The constant $q$ can be put outside the gradient sign. Doing this and then cancelling $q$, we arrive at the formula

$$\boldsymbol{E} = -\nabla\varphi \qquad (8.41)$$

establishing the relation between the field strength and potential.

Taking into account the definition of the gradient, we can write that

$$\boldsymbol{E} = -\frac{\partial\varphi}{\partial x}\boldsymbol{e}_x - \frac{\partial\varphi}{\partial y}\boldsymbol{e}_y - \frac{\partial\varphi}{\partial z}\boldsymbol{e}_z \qquad (8.42)$$

Hence, Eq. (8.41) has the following form in projections onto the coordinate axes:

$$E_x = -\frac{\partial \varphi}{\partial x}, E_y = -\frac{\partial \varphi}{\partial y}, E_z = -\frac{\partial \varphi}{\partial z} \tag{8.43}$$

Similarly, the projection of the vector $\boldsymbol{E}$ onto an arbitrary direction $l$ equals the derivative of $\varphi$ with respect to $l$ taken with the opposite sign, i. e. the rate of diminishing of the potential when moving along the direction $l$:

$$E_l = -\frac{\partial \varphi}{\partial l} \tag{8.44}$$

It is easy to see that Eq. (8.44) is correct by choosing $l$ as one of the coordinate axes and taking Eq. (8.43) into account.

Let us explain Eq (8.41) using as an example the field of a point charge. The potential of this field is expressed by Eq. (8.26). Passing over to Cartesian coordinates, we get the expression

$$\varphi = \frac{1}{4\pi\varepsilon_0}\frac{q}{r} = \frac{1}{4\pi\varepsilon_0}\frac{q}{\sqrt{x^2 + y^2 + z^2}}$$

The partial derivative of this function with respect to $x$ is

$$\frac{\partial \varphi}{\partial x} = -\frac{1}{4\pi\varepsilon_0}\frac{qx}{(x^2 + y^2 + z^2)^{3/2}} = -\frac{1}{4\pi\varepsilon_0}\frac{qx}{r^3}$$

Similarly

$$\frac{\partial \varphi}{\partial y} = -\frac{1}{4\pi\varepsilon_0}\frac{qy}{r^3}, \frac{\partial \varphi}{\partial z} = -\frac{1}{4\pi\varepsilon_0}\frac{qz}{r^3}$$

Using the found values of the derivatives in Eq. (8.42), we arrive at the expression

$$\boldsymbol{E} = \frac{1}{4\pi\varepsilon_0}\frac{q(x\mathrm{e}_x + y\mathrm{e}_y + z\mathrm{e}_z)}{r^3} = \frac{1}{4\pi\varepsilon_0}\frac{q\boldsymbol{r}}{r^3} = \frac{1}{4\pi\varepsilon_0}\frac{q}{r^2}\mathrm{e}_r$$

that coincides with Eq. (8.15).

Equation (8.41) allows us to find the field strength at every point from the known values of $\varphi$ We can also solve the reverse problem i. e find the potential difference between two arbitrary points of a field according to the given values of $\boldsymbol{E}$. For this purpose, we shall take advantage of the circumstance that the work done by the forces of a field on the charge $q$ when it is moved from point 1 to point 2 can be calculated as

$$A_{12} = \int_1^2 q\boldsymbol{E}\mathrm{d}\boldsymbol{l}$$

At the same time in accordance with Eq. (8.30), this work can be written as

$$A_{12} = q(\varphi_1 - \varphi_2)$$

Equating these two expressions and cancelling $q$, we obtain

$$\varphi_1 - \varphi_2 = \int_1^2 \boldsymbol{E}\mathrm{d}\boldsymbol{l} \tag{8.45}$$

The integral can be taken along any line joining points 1 and 2 because the work of the field forces is independent of the path. For circumvention along a closed contour, $\varphi_1 = \varphi_2$ and Eq. (8.45) becomes

$$\oint \boldsymbol{E}\mathrm{d}\boldsymbol{l} = 0 \tag{8.46}$$

(the circle on the integral sign indicates that integration is performed over a closed contour). It roust be noted that this relation holds only for an electrostatic field. We shall see on a later page that the field of moving charges (i.e a field changing with time) is not a potential one. Therefore, condition (8.46) is not observed for it.

An imaginary surface all of whose points have the same potential is called an equipotential surface. Its equation has the form

$$\varphi(x, y, z) = \text{const}$$

The potential does not change in movement along an equipotential surface over the distance $\mathrm{d}\boldsymbol{l}$ ($\mathrm{d}\varphi = 0$) Hence, according to Eq. (8.44), the tangential component of the vector $\boldsymbol{E}$ to the surface equals zero. We thus conclude that the vector $\boldsymbol{E}$ at every point is directed along a normal to the equipotential surface passing through the given point. Bearing in mind that the vector $\boldsymbol{E}$ is directed along a tangent to an $\boldsymbol{E}$ line, we can easily see that the field lines at every point are orthogonal to the equipotential surfaces.

An equipotential surface can be drawn through any point of a field. Consequently, we can construct an infinitely great number of such surfaces. They are conventionally drawn so that the potential difference for two adjacent surfaces is the same everywhere. Thus, the density of the equipotential surfaces allows us to assess the magnitude of the field strength. Indeed, the denser are the equipotential surfaces, the more rapidly does the potential change when moving along a normal to the surface. Hence, $\nabla\varphi$ is greater at the given place, and, therefore, $\boldsymbol{E}$ is greater too.

Figure 8.8 shows equipotential surfaces (more exactly, their intersections with the plane of the drawing) for the field of a point charge. In accordance with the nature of the dependence of $E$ on $r$, equipotential surfaces become the denser, the nearer we approach a charge.

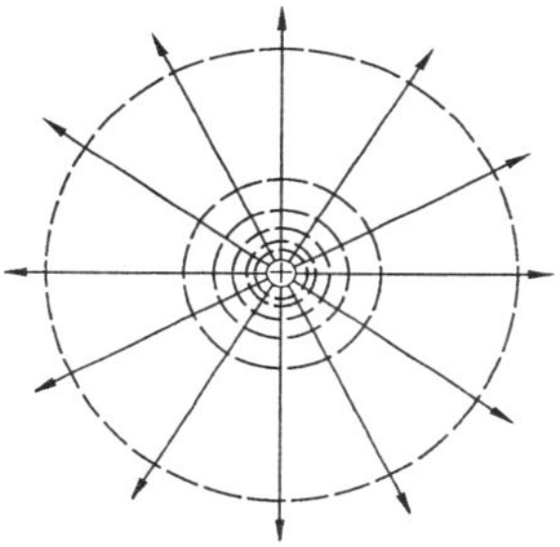

**Fig.8.8**

Equipotential surfaces for a homogeneous field are a collection of equispaced planes at right angles to the direction of the field.

## 8.9 DIPOLE

An electric dipole is defined as a system of two point charges $+q$ and $-q$ identical in value and opposite in sign, the distance between which is much smaller than that to the points at which the field of the system is being determined. The straight line passing through both charges is called the dipole axis.

Let us first calculate the potential and then the field strength of a dipole. This field has axial symmetry. Therefore, the pattern of the field in any plane passing through the dipole axis will be the same, the vector $\boldsymbol{E}$ being in this plane. The position of a point relative to the dipole will be characterized with the aid of the position vector $\boldsymbol{r}$ or with the aid of the polar coordinates $r$ and $\theta$ (Fig. 8.9). We shall introduce the vector $\boldsymbol{l}$ passing from the negative charge to the positive one. The position of the charge $+q$ relative to the center of the dipole is determined by the vector $\boldsymbol{a}$,

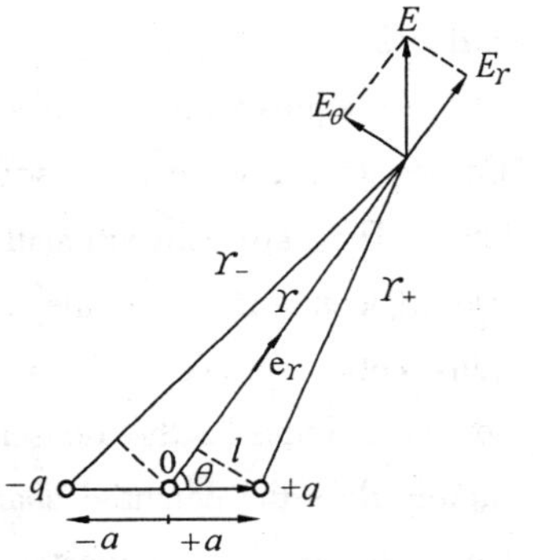


Fig.8.9

and of the charge $-q$ by the vector $-a$. It is obvious that $\mathbf{1} = 2\boldsymbol{a}$ We shall designate the distances to a given point from the charges $+q$ and $-q$ by $r_+$ and $r_-$, respectively.

Owing to the smallness of a in comparison with $r$, we can assume approximately that

$$
\begin{aligned}
r_+ &= r - a\cos\theta = r - \boldsymbol{a}\boldsymbol{e}_r \\
r_- &= r + a\cos\theta = r + \boldsymbol{a}\boldsymbol{e}_r
\end{aligned}
\tag{8.47}
$$

The potential at a point determined by the position vector $\boldsymbol{r}$ is

$$\varphi(r) = \frac{1}{4\pi\varepsilon}\left(\frac{q}{r_+} - \frac{q}{r_-}\right) = \frac{1}{4\pi\varepsilon_0}\frac{q(r_- - r_+)}{r_+ r_-}$$

The product $r_+ r_-$ can be replaced with $r^2$. The difference $r_- - r_+$, according to Eq. (8.47), is $2\boldsymbol{a}\boldsymbol{e}_r = \boldsymbol{l}\boldsymbol{e}_r$, Hence,

$$\varphi(\boldsymbol{r}) = \frac{1}{4\pi\varepsilon_0}\frac{q\boldsymbol{l}\boldsymbol{e}_r}{r^2} = \frac{1}{4\pi\varepsilon_0}\frac{\boldsymbol{p}\boldsymbol{e}_r}{r^2} \tag{8.48}$$

where

$$\boldsymbol{p} = q\boldsymbol{l} \tag{8.49}$$

is a characteristic of a dipole called its electric moment. The vector $\boldsymbol{p}$ is directed along the dipole axis from the negative charge to the posi-

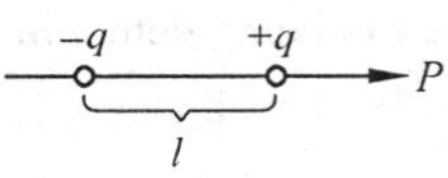


Fig.8.10

tive one (Fig. 8.10).

A glance at Eq. (8.48) shows that the field of a dipole is determined by its electric moment $\boldsymbol{p}$. We shall see below that the behavior of a dipole in an external electric field is also determined by its electric moment $\boldsymbol{p}$. A comparison with Eq. (8.26) shows that the potential of a dipole field diminishes with the distance more rapidly (as $1/r^2$) than the potential of a point charge field (which changes in proportion to $1/r$)

It can be seen from Fig.8.9 that $\boldsymbol{p}\boldsymbol{e}_r = p\cos\theta$. Therefore, Eq.(8.48) can be written as follows:

$$\varphi(r,\theta) = \frac{1}{4\pi\varepsilon_0}\frac{p\cos\theta}{r^2} \tag{8.50}$$

To find the field strength of a dipole, let us calculate the projections of the vector $\boldsymbol{E}$ onto two mutually perpendicular directions by Eq (8.44). One of them is determined by the motion of a point clue to the change in the distance $r$ (with $\theta$ fixed), the other by the motion of the point due to the change in the angle $\theta$ (with $r$ fixed see Fig 8.9). The first projection is obtained by differentiation of Eq.(8.50) with respect to $r$:

$$E_r = -\frac{\partial\varphi}{\partial r} = \frac{1}{4\pi\varepsilon_0}\frac{2p\cos\theta}{r^3} \tag{8.51}$$

We shall find the second projection (let us designate it by $E_\theta$) by taking the ratio of the increment of the potential $\varphi$ obtained when the angle $\theta$ grows by $\mathrm{d}\theta$ to the distance $r\mathrm{d}\theta$ over which the end of the segment $r$ moves [in this case the quantity $\mathrm{d}l$ in Eq(8.44) equals $r\mathrm{d}\theta$]. Thus,

$$E_\theta = -\frac{\partial\varphi}{r\mathrm{d}\theta} = -\frac{1}{r}\frac{\partial\varphi}{\partial\theta}$$

Introducing the value of the derivative of function (8.50) with respect to $\theta$, we get

$$E_\theta = \frac{1}{4\pi\varepsilon_0}\frac{p\sin\theta}{r^3} \tag{8.52}$$

The sum of the squares of Eqs. (8.51) and (8.52) gives the square of the vector $\boldsymbol{E}$ (see Fig. 8.9):

$$E^2 = E_r^2 + E_\theta^2 = (\frac{1}{4\pi\varepsilon_0})^2(\frac{p^2}{r^3})(4\cos^2\theta + \sin^2\theta) =$$

$$(\frac{1}{4\pi\varepsilon_0})^2(\frac{p}{r^3})^2(1 + 3\cos^2\theta)$$

Hence

$$E = \frac{1}{4\pi\varepsilon_0}\frac{p}{r^3}\sqrt{1 + 3\cos^2\theta} \qquad (8.53)$$

Assuming in Eq. (8.53) that $\theta = 0$, we get the strength on the dipole axis:

$$\boldsymbol{E}_{//} = \frac{1}{4\pi\varepsilon_0}\frac{2p}{r^3} \qquad (8.54)$$

The vector $\boldsymbol{E}_{//}$ is directed along the dipole axis. This is in agreement with the axial symmetry of the problem. Examination of Eq. (8.51) shows that $E_r > 0$ when $\theta = 0$, and $E_r > 0$ when $\theta = \pi$. This signifies that in any case the vector $\boldsymbol{E}_{//}$ has a direction coinciding with that from $-q$ to $+q$ (i.e. with the direction of $\boldsymbol{p}$). Equation (8.54) can therefore be written in the vector form:

$$\boldsymbol{E}_{//} = \frac{1}{4\pi\varepsilon_0}\frac{2\boldsymbol{p}}{r^3} \qquad (8.55)$$

Assuming in Eq. (8.53) that $\theta = \pi/2$, we get the field strength on the straight line passing through the center of the dipole and perpendicular to its axis:

$$\boldsymbol{E}_{\perp} = \frac{1}{4\pi\varepsilon_0}\frac{p}{r^3} \qquad (8.56)$$

By Fq. (8.51), when $\theta = \pi/2$, the projection $E_r$ equals zero. Hence, the vector $\boldsymbol{E}_{\perp}$ is parallel to the dipole axis. It follows from Eq. (8.52) that when $\theta = \pi/2$, the projection $E_\theta$ is positive. This signifies that the vector $\boldsymbol{E}_{\perp}$ is directed toward the growth of the angle $\theta$ i.e. antiparallel to the vector $\boldsymbol{p}$.

The field strength of a dipole is characterized by the circumstance that it diminishes with the distance from the dipole in proportion to $1/r^3$, i.e. more rapidly than the field strength of a point charge (which diminishes in proportion to $1/r^2$).

Figure 8.11 shows $\boldsymbol{E}$ lines (the solid lines) and equipotential surfaces (the dash lines) of the field of a dipole. According to Eq. (8.50), when $\theta = \pi/2$, the potential vanishes for all the $r$'s. Thus, all the points of a plane at right angles to the dipole axis and passing through its middle have a zero potential. This could have been predicted because the distances from the charges $+q$ and $-q$ to any point of this plane are identical.

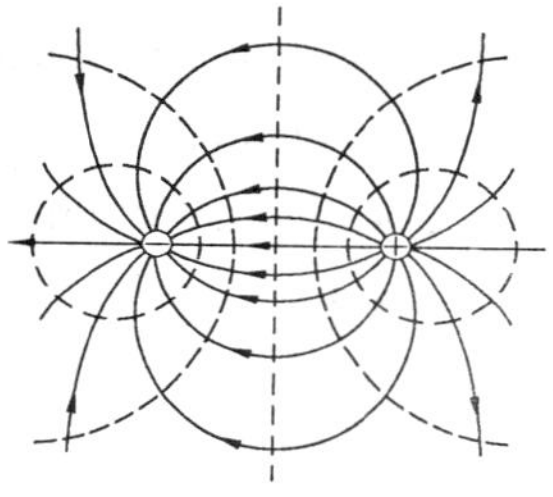

**Fig.8.11**

Now let us turn to the behavior of a dipole in an external electric field. If a dipole is placed in a homogeneous electric field, the charges $+q$ and $-q$ forming the dipole will be under the action of the forces $\boldsymbol{F}_1$ and $\boldsymbol{F}_2$ equal in magnitude, but opposite in direction (Fig.8.12). These forces form a couple whose arm is $l \sin \alpha$, i.e. depends on the orientation of the dipole relative to the field. The magnitude of each of the forces is $qE$. Multiplying it by the arm, we get the magnitude of the torque acting on a dipole:

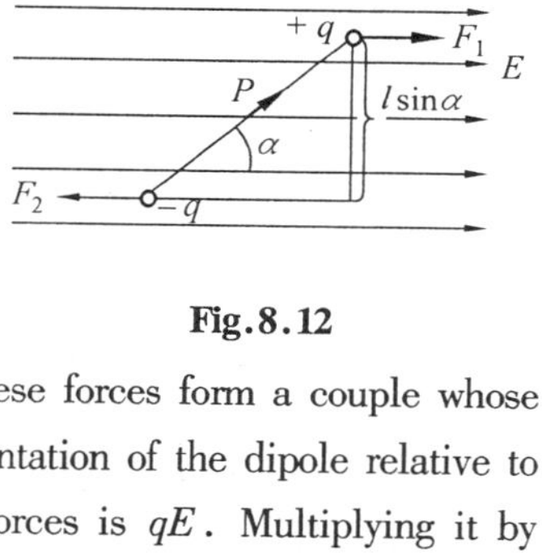


**Fig.8.12**

$$T = qEl\sin \alpha = pE\sin \alpha \tag{8.57}$$

($p$ is the electric moment of the dipole). It is easy to see that Eq. (8.57) can be written in the vector form

$$\boldsymbol{T} = [\boldsymbol{pE}] \tag{8.58}$$

The torque (8.58) tends to turn a dipole so that its electric moment $\boldsymbol{p}$ is in the direction of the field.

Let us find the potential energy belonging to a dipole in an external electric field. By Eq. (8.29), this energy is

$$W_p = q\varphi_+ - q\varphi_- = q(\varphi_+ - \varphi_-) \tag{8.59}$$

Here $\varphi_+$ and $\varphi_-$ are the values of the potential of the external field at the points where the charges $+q$ and $-q$ are placed.

The potential of a homogeneous field diminishes linearly in the direction of the vector $\boldsymbol{E}$. Assuming that the $x$-axis is this direction (Fig. 8.13), we can write that $E = E_x = -\mathrm{d}\varphi/\mathrm{d}x$. A glance at Fig. 8.13 shows that the difference $\varphi_+ - \varphi_-$ equals the increment of the potential on the segment $\Delta x = l\cos\alpha$

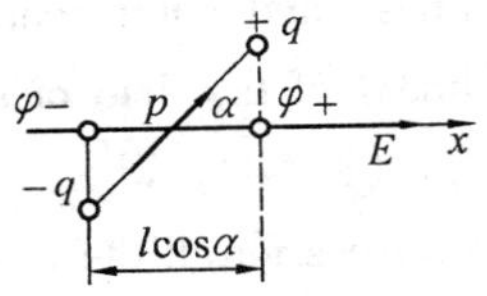


Fig.8.13

$$\varphi_+ - \varphi_- = \frac{\mathrm{d}\varphi}{\mathrm{d}x} l\cos\alpha = -El\cos\alpha$$

Introducing this value into Eq. (8.59). we find that

$$W_p = -qEl\cos\alpha = -pE\cos\alpha \tag{8.60}$$

Here $\alpha$ is the angle between the vectors $\boldsymbol{p}$ and $\boldsymbol{E}$. we can therefore write Eq. (8.60) in the form

$$W_p = -\boldsymbol{pE} \tag{8.61}$$

We must note that this expression takes no account of the energy of interaction of the charges $+q$ and $-q$ forming a dipole.

We have obtained Eq. (8.61) assuming for simplicity's sake that the field is homogeneous. This equation also holds, however, for an inhomogeneous field.

Let us consider a dipole in an inhomogeneous field that is symmetrical relative to the $x$-axis. Let the center of the dipole be on this axis, the dipole electric moment making with the axis an angle $\alpha$, differing from $\pi/2$ (Fig. 8.14).

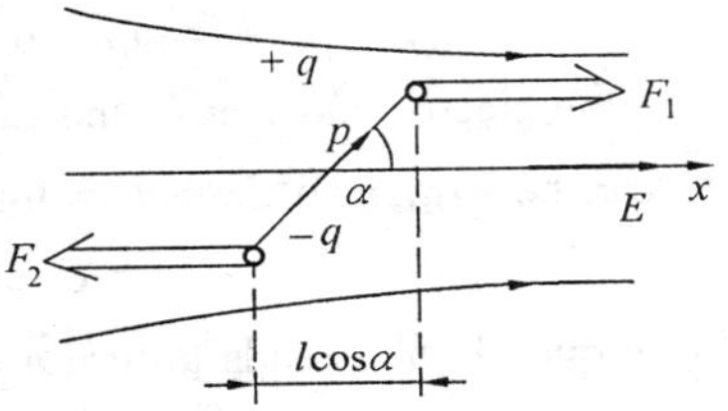


Fig.8.14

In this case, the forces acting on the dipole charges are not identical in magnitude. Therefore, apart from the rotational moment (torque), the dipole will experience a force tending to move it in the direction of the $x$-axis. To find the value of this force, we shall use Eq.(8.40), according to which

$$F_x = -\partial W_p/\partial x, F_y = -\partial W_p/\partial y, F_z = -\partial W_p/\partial z$$

In view of Eq.(8.60), we can write

$$W_p(x,y,z) = -pE(x,y,z)\cos\alpha$$

(we consider the orientation of the dipole relative to the vector $\boldsymbol{E}$ to be constant, $\alpha = \text{const}$).

For points on the $x$-axis, the derivatives of $E$ with respect to $y$ and $z$ are zero. accordingly, $\partial W_p/\partial y = \partial W_p/\partial z = 0$. Thus, only the force component $F_x$ differs from zero. It is

$$F_x = -\frac{\partial W_p}{\partial x} = p\frac{\partial E}{\partial x}\cos\alpha \tag{8.62}$$

This result can be obtained if we take account of the fact that the field strength at the points where the charges $+q$ and $-q$ are (see Fig.8.14) differs by the amount $(\partial E/\partial x)l\cos\alpha$. Accordingly, the difference between the forces acting on the charges is $q(\partial E/\partial x)l\cos\alpha$, which coincides with Eq.(8.62).

When $\alpha$ is less than $\pi/2$, the value of $F_x$ determined by Eq. (8.62) is positive. This signifies that under the action of the force the dipole is pulled into the region of a stronger field (see Fig.8.14). When $\alpha$ is greater than $\pi/2$, the dipole is pushed out of the field. In the case shown in Fig.8.15, only the derivative $\partial E/\partial y$ differs from zero for points on the $y$-axis. Therefore, the force acting on the dipole is determined by the component

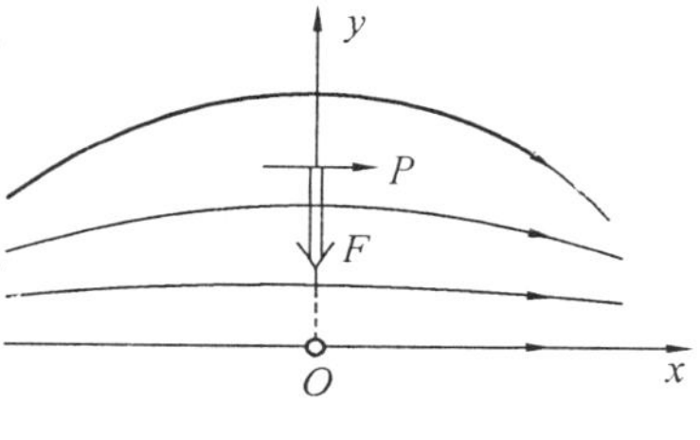


**Fig.8.15**

$$F_y = -\frac{\partial W_p}{\partial y} = p\frac{\partial E}{\partial y}(\cos\alpha = 1)$$

The derivative $\partial E/\partial y$ is negative. Consequently, the force is directed as shown in the figure. Thus, in this case too, the dipole is pulled into the field.

We shall note that like $-\partial W_p/\partial x$ gives the projection of the force acting on the system onto the $x$-axis, so does the derivative of Eq. (8.60) with respect to a taken with the opposite sign give the projection of the torque onto the $\alpha$ – "axis": $T_\alpha = -pE\sin\alpha$. The minus sign was obtained because the a – "axis" and the torque $T$ are directed oppositely (see Fig.8.12).

## The Further Study:

### A Description of The Properties of Vector Fields

To continue our study of the electric field, we must acquaint ourselves with the mathematical tools used to describe the properties of vector fields. These tools are called vector analysis. In the present section, we shall treat the fundamental concepts and selected formulas of vector analysis, and also prove its two main theorems—the Ostrogradsky – Gauss theorem (sometimes called Gauss's divergence theorem) and Stokes's theorem.

The quantities used in vector analysis can be best illustrated for the field of the velocity vector of a flowing liquid. We shall therefore introduce these quantities while dealing with the flow of an ideal incompressible liquid, and then extend the results obtained to vector fields of any nature.

We are already acquainted with one of the concepts of vector analysis. This is the gradient, used to characterize scalar fields. If the value of the scalar quantity $\varphi = \varphi(x, y, z)$ is compared with every point $P$ having the coordinates $x$, $y$, $z$, we say that the scalar field of $\varphi$ has been set. The gradient of the quantity $\varphi$ is defined as the vector

$$\text{grad } \varphi = \frac{\partial \varphi}{\partial x} \mathrm{e}_x + \frac{\partial \varphi}{\partial y} \mathrm{e}_y + \frac{\partial \varphi}{\partial z} \mathrm{e}_z \tag{8.63}$$

The increment of the function $\varphi$ upon displacement over the length $\mathrm{d}\boldsymbol{l} = \boldsymbol{e}_x \mathrm{d}x + \boldsymbol{e}_y \mathrm{d}y + \boldsymbol{e}_z \mathrm{d}z$ is

$$\mathrm{d}\varphi = \frac{\partial \varphi}{\partial x} \mathrm{d}x + \frac{\partial \varphi}{\partial y} \mathrm{d}y + \frac{\partial \varphi}{\partial z} \mathrm{d}z$$

which can be written in the form

$$\mathrm{d}\varphi = \nabla \varphi \cdot \mathrm{d}\boldsymbol{l} \tag{8.64}$$

Now we shall go over to establishing the characteristics of vector fields.

## Vector Flux

Assume that the flow of a liquid is characterized by the field of the velocity vector. The volume of liquid flowing in unit time through an imaginary surface $S$ is called the flux of the liquid through this surface. To find the flux, let us divide the surface into elementary sections of the size $\Delta S$. It can be seen from Fig. 8.16 that during the time $\Delta t$ a volume of liquid equal to

$$\Delta V = (\Delta S \cos \alpha) v \Delta t$$

will pass through section $\Delta S$. Dividing this volume by the time $\Delta t$, we shall find the flux through surface $\Delta S$:

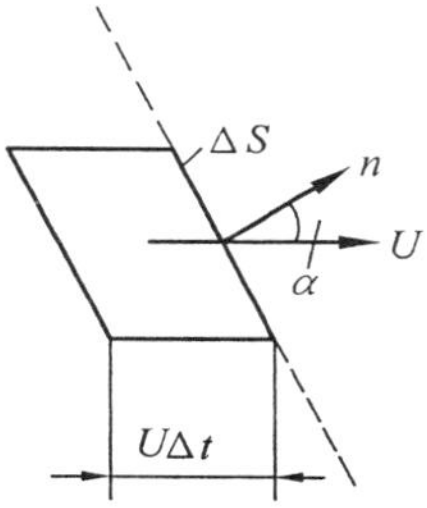


**Fig.8.16**

$$\Delta \varphi = \frac{\Delta V}{\Delta t} = \Delta S v \cos \alpha$$

Passing over to differentials, we find that

$$\mathrm{d}\Phi = (v \cos \alpha) \mathrm{d}S \tag{8.65}$$

Equation (8.65) can be written in two other ways. First, if we take into account that $v\cos\alpha$ gives the projection of the velocity vector onto the normal $n$ to area $\mathrm{d}S$, we can write Eq. (8.65) in the form

$$d\Phi = v_n dS \quad (8.66)$$

Second, we can introduce the vector $d\boldsymbol{S}$ whose magnitude equals that of area $dS$, while its direction coincides with the direction of a normal $n$ to the area:

$$d\boldsymbol{S} = dS \cdot \boldsymbol{n}$$

Since the direction of the vector $\boldsymbol{n}$ is chosen arbitrarily (it can be directed to either side of the area), then $d\boldsymbol{S}$ is not a true vector, but is a pseudo vector. The angle $a$ in Eq. (8.65) is the angle between the vectors $\boldsymbol{v}$ and $d\boldsymbol{S}$. Hence, this equation can be written in the form

$$d\Phi = \boldsymbol{v} d\boldsymbol{S} \quad (8.67)$$

By summating the fluxes through all the elementary areas into which we have divided surface $S$, we get the flux of the liquid through $S$:

$$\Phi_v = \int_S \boldsymbol{v} d\boldsymbol{S} = \int_S v_n dS \quad (8.68)$$

A similar expression written for an arbitrary vector field $\boldsymbol{a}$, i.e. the quantity

$$\Phi_a = \int_S \boldsymbol{a} d\boldsymbol{S} = \int_S a_n dS \quad (8.69)$$

is called the flux of the vector $\boldsymbol{a}$ through surface $\boldsymbol{S}$. In accordance with this definition, the flux of a liquid can be called the flux of the vector $\boldsymbol{v}$ through the relevant surface [see Eq. (8.68)].

The flux of a vector is an algebraic quantity. Its sign depends on the choice of the direction of a normal to the elementary areas into which surface $S$ is divided in calculating the flux. Reversal of the direction of the normal changes the sign of $a_n$ and, therefore, the sign of the quantity (8.69). The customary practice for closed surfaces is calculation of the flux "emerging outward" from the region enclosed by the surface. Accordingly, in the following we shall always implicate that $\boldsymbol{n}$ is an outward normal.

We can give an illustrative geometrical interpretation of the vector flux. For this purpose, we shall represent a vector field by a system of lines a constructed so that the density of the lines at every point is numerically equal to the magnitude of the vector ***a*** at the same point of the field. Let us find the number $\Delta N$ of intersections of the field lines with the imaginary area $\Delta S$. A glance at Fig. 8.17 shows that this number equals the density of the lines (i.e. a) multiplied by $\Delta S_{\perp} = \Delta s \cos \alpha$:

$$\Delta N = \alpha \Delta S \cos \alpha = a_n \Delta S$$

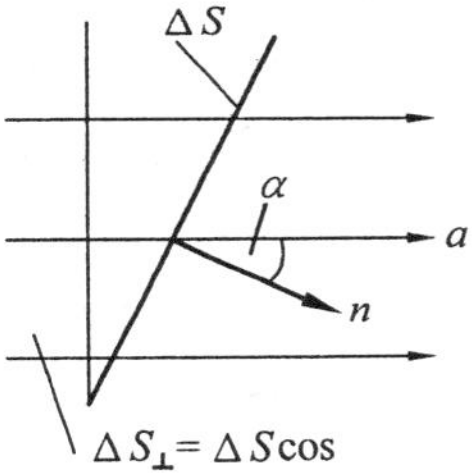


**Fig. 8.17**

We are speaking only about the numerical equality between $\Delta N$ and $a_n \Delta S$ This is why the equality sign is confined in parentheses. According to Eq. (8.69), the expression $a_n \Delta S$ is $\Delta \Phi_a$ – the flux of the vector ***a*** through area $\Delta S$. Thus,

$$\Delta N (=) \Delta \Phi_a \tag{8.70}$$

For the sign of $\Delta N$ to coincide with that of $\Delta \Phi_a$, we must consider those intersections to be positive for which the angle $\alpha$ between the positive direction of a field line and a normal to the area is acute. The intersection should be considered negative if the angle $\alpha$ is obtuse. all three intersections are positive: $\Delta N = +3$ ($\Delta \Phi_a$ in tills case is also positive because $a_n > 0$). If the direction of the normal in Fig. 8.17 is reversed, the intersections will become negative ($\Delta N = -3$), and the flux $\Delta \Phi_a$ will also be negative.

Summation of Eq. (8.70) over the finite imaginary surface $S$ yields

$$\Phi_a = \sum \Delta N = N_+ - N_- \tag{8.71}$$

where $N_+$ and $N_-$ are the total number of positive and negative intersections of the field lines with surface $S$, respectively.

The reader may be puzzled by the circumstance that since the flux, as a rule, is expressed by a fractional number, the number of intersections of the field lines with a surface compared with the flux will also be fractional. Do not be confused by this, however. Field lines are a purely conditional image deprived of a physical meaning.

Let us take an imaginary surface in the form of a strip of paper whose bottom part is twisted relative to the top one through the angle $\pi$(Fig.8.18). The direction of a normal must be chosen identically for the entire surface. Hence, if in the top part of the strip a positive normal is directed to the right, then in the bottom part a normal will be directed to the left. Accordingly, the intersections of the field lines depicted in Fig.8.18 with the top half of the surface must be considered positive, and with the bottom half, negative.

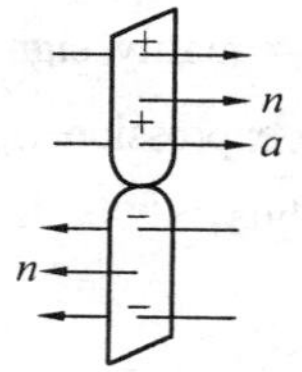


**Fig.8.18**

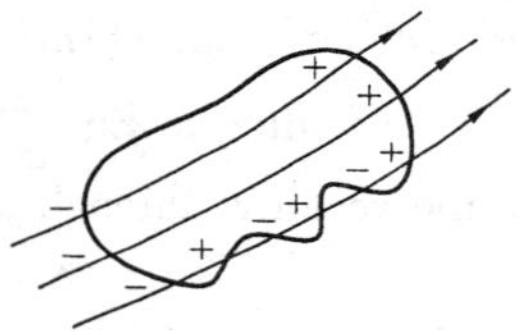

**Fig.8.19**

An outward normal is considered to be positive for a closed surface (Fig.8.19). Therefore, the intersections corresponding to outward protrusion of the lines (in this case the angle $\alpha$ is acute) must be taken with the plus sign, and the ones appearing when the lines enter the surface (in this case the angle $\alpha$ is obtuse) must be taken with the minus sign.

Inspection of Fig.8.19 shows that when the field lines enter a closed surface continuously, each line when intersecting the surface enters it and emerges from it the same number of times. As a result, the flux of the corresponding vector through this surface equals zero. It is easy to see that if field lines end inside a surface, the vector flux through the closed surface will numerically equal the difference between the number of lines beginning inside the surface ($N_{beg}$) and the number of lines terminating inside the surface ($N_{term}$):

$$\Phi_a (=) N_{\text{beg}} - N_{\text{term}} \tag{8.72}$$

The sign of the flux depends on which of these numbers is greater. When $N_{\text{beg}} = N_{\text{term}}$, the flux equals zero.

Divergence. Assume that we are given the field of the velocity vector of an incompressible continuous liquid. Let us take an imaginary closed surface $S$ in the vicinity of point $P$ (Fig. 8.20). If in the volume confined by this surface no liquid appears and no liquid vanishes, then the flux flowing outward through the surface will evidently equal zero. A liquid flux $\Phi_v$ other than zero will indicate that there are liquid sources or sinks inside the surface, i.e. points at which the liquid enters the volume (sources) or emerges from it (sinks). The magnitude of the flux determines the total algebraic power of the sources and sinks. When the sources predominate over the sinks, the magnitude of the flux will be positive, and when the sinks predominate, negative.

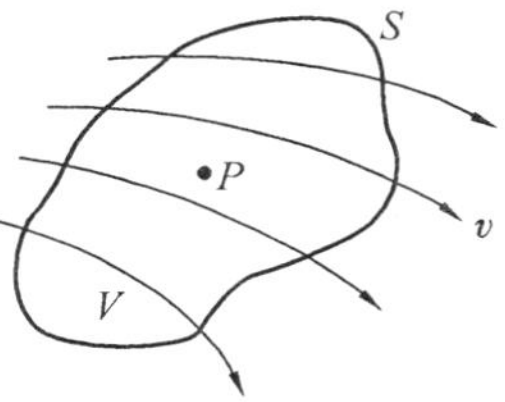


**Fig.8.20**

The quotient obtained when dividing the flux $\Phi_v$ by the volume which it flows out from, i.e

$$\frac{\Phi_v}{V} \tag{8.73}$$

gives the average unit power of the sources confined in the volume $V$. In the limit when $V$ tends to zero, i.e. when the volume $V$ contracts to point $P$, expression (8.73) gives the true unit power of the sources at point $P$, which is called the divergence of the vector $\boldsymbol{v}$ (it is designated by div $\boldsymbol{v}$). Thus, by definition,

$$\text{div}\ \boldsymbol{v} = \lim_{V \to p} \frac{\Phi}{V}$$

The divergence of any vector $\boldsymbol{a}$ is determined in a similar way:

$$\text{div}\boldsymbol{a} = \lim_{V \to p} \frac{\Phi_a}{V} = \lim_{V \to p} \frac{1}{V} \oint \boldsymbol{a} \, \mathrm{d}\boldsymbol{S} \tag{8.74}$$

The integral is taken over arbitrary closed surface $S$ surrounding point $P$; $V$ is the volume confined by this surface. Since the transition $V \to P$ is being performed upon which $S$ tends to zero, we can assume that Eq. (8.74) cannot depend on the shape of the surface. This assumption is confirmed by strict calculations.

Let us surround point $P$ with a spherical surface of an extremely small radius $r$ (Fig.8.21). Owing to the smallness of $r$, the volume $V$ enclosed by the sphere will also be very small. We can therefore consider with a high degree of accuracy that the value of div $\boldsymbol{a}$ within the limits of the volume $V$ is constant. In this case, we can write in accordance with Eq.(8.74) that

$$\Phi_a \approx \operatorname{div} \boldsymbol{a} \cdot V$$

where $\Phi_a$ is the flux of the vector $\boldsymbol{a}$ through the surface surrounding the volume $V$. By Eq. (8.72), $\Phi_a$ equals $N_{\text{beg}}$, the number of lines of a beginning inside $V$ if div $\boldsymbol{a}$ at point $P$ is positive, or $N_{\text{term}}$ the number of lines of a terminating inside $V$ if div $\boldsymbol{a}$ at point $P$ is negative.

It follows from the above that the lines of the vector $\boldsymbol{a}$ begin in the closest vicinity of a point with a positive divergence. The field lines "diverge" from this point; the latter is the "source" of the field (Fig. 8.21a). On the other hand, in the vicinity of a point with a negative divergence, the lines of the vector $\boldsymbol{a}$ terminate. The field lines "converge" toward this point; the latter is the "sink" of the field (Fig. 8.21b). The greater the absolute value of div $\boldsymbol{a}$, the bigger is the number of lines that begin or terminate in the vicinity of the given point.

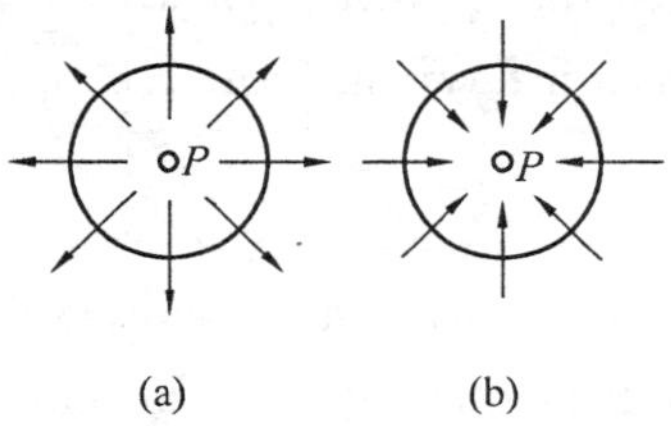


**Fig.8.21**

It can be seen from definition (8.74) that the divergence is a scalar function of the coordinates determining the positions of points in space (briefly—a point function). Definition (8.74) is the most general one that is independent of the kind of coordinate system used.

Let us find an expression for the divergence in a Cartesian coordinate system. We shall consider a small volume in the form of a parallelepiped with ribs parallel to the coordinate axes in the vicinity of point $P(x, y, z)$ (Fig. 8.22). The vector flux through the surface of the parallelepiped is formed from the fluxes passing through each of the six faces separately.

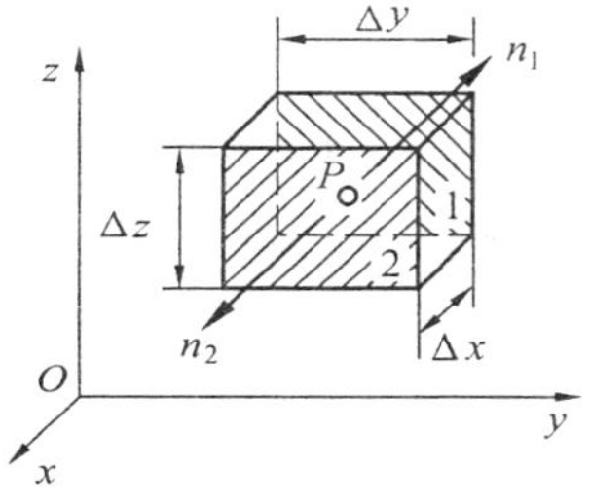


**Fig.8.22**

Let us find the flux through the pair of faces perpendicular to the $x$-axis (in Fig.8.22 these faces are designated by diagonal hatching and by the numbers 1 and 2). The outward normal $\boldsymbol{n}_2$ to face 2 coincides with the direction of the $x$-axis. Hence, for points of this face, $a_{n_2} = a_x$. The outward normal $\boldsymbol{n}_1$ to face 1 is directed oppositely to the $x$-axis. Therefore, for points on this face, $a_{n_1} = -a_x$. The flux through face 2 can be written in the form

$$a_{x,2}\Delta y\Delta z$$

where $a_{x,2}$ is the value of $a_x$ averaged over face 2. The flux through face 1 is

$$-a_{x,1}\Delta y\Delta z$$

where $a_{x,1}$ is the average value of $a_x$ for face 1. The total flux through faces 1 and 2 is determined by the expression

$$(a_{x,2} - a_{x,1})\Delta y\Delta z \tag{8.75}$$

The difference $a_{x,2} - a_{x,1}$ is the increment of the average (over a face) value of $a_x$ upon a displacement along the $x$-axis by $\Delta x$. Owing to the smallness of the parallelepiped (we remind our reader that we shall let its dimensions shrink to zero), this increment can be written in the form

$(\partial a_x/\partial x)\Delta x$, where the value $\partial a_x/\partial x$ is taken at point $P$. Therefore, Eq. (8.75) becomes

$$\frac{\partial a_x}{\partial x}\Delta x\Delta y\Delta z = \frac{\partial a_x}{\partial x}\Delta V$$

Similar reasoning allows us to obtain the following expressions for the fluxes through the pairs of faces perpendicular to the $y$- and $z$-axes:

$$\frac{\partial a_y}{\partial y}\Delta V \quad \text{and} \quad \frac{\partial a_z}{\partial z}\Delta V$$

Thus, the total flux through the entire close surface is determined by the expression

$$\Phi_a = \left(\frac{\partial a_x}{\partial x} + \frac{\partial a_y}{\partial y} + \frac{\partial a_z}{\partial z}\right)\Delta V$$

Dividing this expression by $\Delta V$, we shall find the divergence of the vector $\boldsymbol{a}$ at point $P(x, y, z)$:

$$\operatorname{div} \boldsymbol{a} = \frac{\partial a_x}{\partial x} + \frac{\partial a_y}{\partial y} + \frac{\partial a_z}{\partial z} \tag{8.76}$$

## The Oslrogradsky – Gauss Theorem

If we know the divergence of the vector $\boldsymbol{a}$ at every point of space, we can calculate the flux of this vector through any closed surface of finite dimensions. Let us first do this for the flux of the vector $\boldsymbol{v}$ (a liquid flux). The product of div $\boldsymbol{v}$ and $\mathrm{d}V$ gives the power of the sources of the liquid confined within the volume $\mathrm{d}V$. The sum of such products, i.e $\int \operatorname{div} v \cdot \mathrm{d}V$, gives the total algebraic power of the sources confined in the volume $V$ over which integration is performed. Owing to incompressibility of the liquid, the total power of the sources must equal the liquid flux emerging through surface $S$ enclosing the volume $V$. We thus arrive at the equation

$$\oint_S \boldsymbol{v}\mathrm{d}\boldsymbol{S} = \int_V \operatorname{div}\boldsymbol{v}\mathrm{d}V$$

A similar equation holds for a vector field of any nature:

$$\oint_S \boldsymbol{a}\,\mathrm{d}\boldsymbol{S} = \int_V \mathrm{div}\boldsymbol{a}\,\mathrm{d}V \tag{8.77}$$

This relation is called the Ostrogradsky – Gauss theorem. The integral in the left-hand side of the equation is calculated over an arbitrary closed surface $S$, and the integral in the right hand side over the volume $V$ enclosed by this surface.

**Circulation**

Let us revert to the flow of an ideal incompressible liquid. Imagine a closed line the contour $\Gamma$. Assume that in some way or other we have instantaneously frozen the liquid in the entire volume except for a very thin closed channel of constant cross section including the contour $\Gamma$ (Fig. 8.23). Depending on the nature of the velocity vector field, the liquid in the channel formed will either be stationary or move along the contour (circulate) in one of the two possible directions. Let us take the quantity equal to the product of the velocity of the liquid in the channel and the length of the contour $l$ as a measure of this motion. This quantity is called the circulation of the vector $\boldsymbol{v}$ around the contour $\Gamma$ Thus,

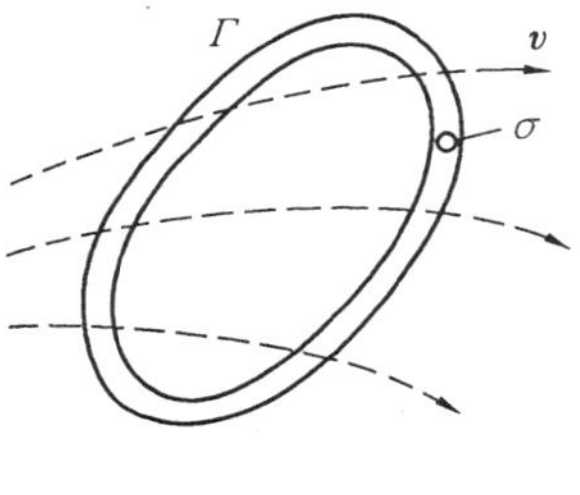


**Fig.8.23**

$$\text{circulation of } \boldsymbol{v} \text{ around } \Gamma = vl$$

(since we assumed that the channel has a constant cross section, the magnitude of the velocity $v = \text{const}$).

At the moment when the walls freeze, the velocity component perpendicular to a wall will be eliminated in each of the liquid particles, and only the velocity component tangent to the contour will remain, i.e. $v_l$. The momentum $\mathrm{d}\boldsymbol{p}_l$ is associated with this component. The magnitude of the momentum for a liquid particle contained within a segment of the channel of length $\mathrm{d}l$ is $\rho\sigma v_l \mathrm{d}l$ ($\rho$ is the density of the liquid, and $\sigma$ is the cross-sectional area of the channel). Since the liquid is ideal, the action of the walls can change only the direction of the vector $\mathrm{d}\boldsymbol{p}_l$, but not its magnitude. The interaction between the liquid particles will cause a redistribution of the momentum between them that will level out the velocities of all the particles. The algebraic sum of the tangential components of the momenta cannot change: the momentum acquired by one of the interacting particles equals the momentum lost by the second particle. This signifies that

$$\rho\sigma v l = \oint_{\Gamma} \rho\sigma v_l \mathrm{d}l$$

where $v$ is the circulation velocity, and $v_l$ is the tangential component of the liquid's velocity in the volume $\sigma \mathrm{d}l$ at the moment of time preceding the freezing of the channel walls. Cancelling $\rho\sigma$, we get

$$\text{circulation of } \boldsymbol{v} \text{ around } \Gamma = vl = \oint v_l \mathrm{d}l$$

The circulation of any vector $\boldsymbol{a}$ around an arbitrary closed contour $T$ is determined in a similar way

$$\text{circulation of } \boldsymbol{a} \text{ around } \Gamma = \oint_{\Gamma} \boldsymbol{a}\mathrm{d}\boldsymbol{l} = \oint_{\Gamma} a_l \mathrm{d}l \qquad (8.78)$$

It may seem that for the circulation to be other than zero the vector lines must be closed or at least bent in some way or other in the direction of circumventing the contour. It is easy to see that this assumption is wrong. Let us consider the laminar flow of water in a river. The velocity of the water directly at the river bottom is zero and grows as we approach the surface of the water (Fig. 8.24). The streamlines (lines of the vector $\boldsymbol{v}$) are straight. Notwithstanding this fact, the circulation of the vector $\boldsymbol{v}$ around the contour depicted by the dash line obviously differs from zero. On the other hand, in a field with curved lines, the circulation may equal zero.

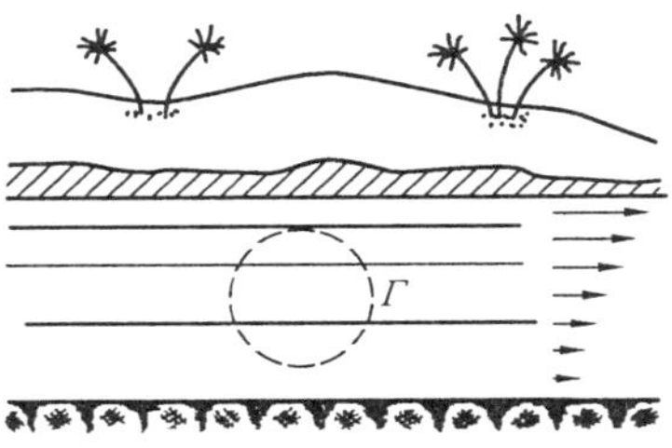


Fig.8.24

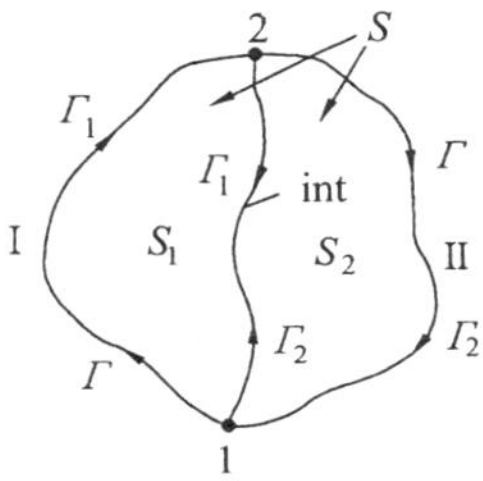


Fig.8.25

Circulation has the property of additivity. This signifies that the sum of the circulations around contours $\Gamma_1$ and $\Gamma_2$ enclosing neighboring surfaces $S_1$ and $S_2$(Fig.8.25) equals the circulation around contour $\Gamma$ enclosing surface $S$, which is the sum of surfaces $S_1$ and $S_2$. Indeed, the circulation $C_1$ around the contour bounding surface $S_1$ can be represented as the sum of the integrals

$$C_1 = \oint_{\Gamma_1} \boldsymbol{a} \mathrm{d}\boldsymbol{l} = \int_1^2 \boldsymbol{a} \mathrm{d}\boldsymbol{l} + \int_2^1 \boldsymbol{a} \mathrm{d}\boldsymbol{l} \tag{8.79}$$

The first integral is taken over section I of the outer contour, the second over the interface between surfaces $S_1$ and $S_2$ in direction 2 – 1.

Similarly, the circulation $C_2$ around the contour enclosing surface $S_2$ is

$$C_2 = \oint_{\Gamma_2} \boldsymbol{a} \mathrm{d}\boldsymbol{l} = \int_2^1 \boldsymbol{a} \mathrm{d}\boldsymbol{l} + \int_1^2 \boldsymbol{a} \mathrm{d}\boldsymbol{l} \tag{8.80}$$

The first integral is taken over section II of the outer contour, the second over the interface between surfaces $S_1$ and $S_2$ in direction 1 – 2.

The circulation around the contour bounding total surface $S$ can be represented in the form

$$C = \oint_{\Gamma} \boldsymbol{a} \mathrm{d}\boldsymbol{l} = \int_1^2 \boldsymbol{a} \mathrm{d}\boldsymbol{l} + \int_2^1 \boldsymbol{a} \mathrm{d}\boldsymbol{l} \tag{8.81}$$

The second addends in Eqs. (8.79) and (8.80) differ only in their sign. Therefore, the sum of these expressions will equal Eq. (8.81).

Thus,

$$C = C_1 + C_2 \tag{8.82}$$

Equation (8.82) which we have proved does not depend on the shape of the surfaces and holds for any number of addends. Hence, if we divide an arbitrary open surface $S$ into a great number of elementary surfaces $\Delta S$ (Fig. 8.26), then the circulation around the contour enclosing $S$ can be written as the sum of the elementary circulations $\Delta C$ around the contours enclosing the $\Delta S$'s:

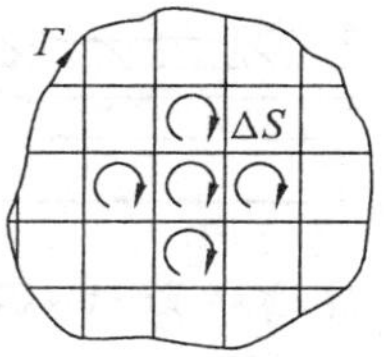

**Fig.8.26**

$$C = \sum \Delta C_i \tag{8.83}$$

### Curl

The additivity of the circulation permits us to introduce the concept of unit circulation, i.e. consider the ratio of the circulation $C$ to the magnitude of surface $S$ around which the circulation "flows". When surface $S$ is finite, the ratio $C/S$ gives the mean value of the unit circulation. This value characterizes the properties of a field averaged over surface $S$. To obtain the characteristic of the field at point $P$, we must reduce the dimensions of the surface, making it shrink to point $P$. The ratio $C/S$ tends to a limit that characterizes the properties of the field at point $P$.

Thus, let us take an imaginary contour $\Gamma$ in a plane passing through point $P$, and consider the expression

$$\lim_{S \to P} \frac{C_a}{S} \tag{8.84}$$

where $C_a$ = circulation of the vector $\boldsymbol{a}$ around the contour $\Gamma$

$S$ = surface area enclosed by the contour.

Limit (8.84) calculated for an arbitrarily oriented plane cannot be an exhaustive characteristic of the field at point $P$ because the magnitude of this limit depends on the orientation of the contour in space in addition to the properties of the field at point $P$. This orientation can be given by the direction of a positive normal $\boldsymbol{n}$ to the plane of the contour (a positive normal is one that is associated with the direction of circumvention of the contour in integration by the right-hand screw rule). In determining limit (8.84) at the same point $P$ for different directions $\boldsymbol{n}$, we shall obtain different values. For opposite directions, these values will differ only in their sign (reversal of the direction $\boldsymbol{n}$ is equivalent to reversing the direction of circumvention of the contour in integration, which only causes a change in the sign of the circulation). For a certain direction of the normal, the magnitude of expression (8.84) at the given point will be maximum.

Thus, quantity (8.84) behaves like the projection of a vector onto the direction of a normal to the plane of the contour around which the circulation is taken. The maximum value of quantity (8.84) determines the magnitude of this vector, and the direction of the positive normal $\boldsymbol{n}$ at which the maximum is reached gives the direction of the vector. This vector is called the curl of the vector $\boldsymbol{a}$. Its symbol is curl $\boldsymbol{a}$. Using this notation, we can write expression (8.84) in the form

$$(\text{curl } a)_n = \lim_{S \to P} \frac{C_a}{S} = \lim_{S \to P} \frac{1}{S} \oint_{\Gamma} \boldsymbol{a} \mathrm{d}\boldsymbol{l} \qquad (8.85)$$

We can obtain a graphical picture of the curl of the vector $\boldsymbol{v}$ by imagining a small and light fan impeller placed at the given point of a flowing liquid (Fig. 8.27). At the spots where the curl differs from zero, the impeller will rotate, its velocity being the higher, the greater in value is the projection of the curl onto the impeller axis.

Equation (8.85) defines the vector curl $\boldsymbol{a}$. This definition is a most general one that does not depend on the kind of coordinate system used. To find expressions for the projections of the vector curl $\boldsymbol{a}$ onto the axes of a Cartesian coordinate system, we must determine the values of quantity

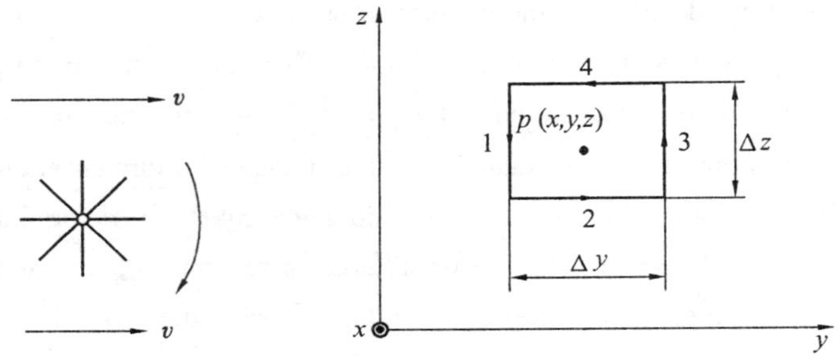


**Fig.8.27**

(8.85) for such orientations of area $S$ for which the normal $\boldsymbol{n}$ to the area coincides with one of the axes $x$, $y$, $z$. If, for example, we direct $\boldsymbol{n}$ along the $x$-axis, then(8.85) becomes (curl$\boldsymbol{a}_x$). Contour $\Gamma$ in this case is arranged in a plane parallel to the coordinate plane $yz$. Let us take this contour in the form of a rectangle with the sides $\Delta y$ and $\Delta z$ (Fig.8. 28 the $x$-axis is directed toward us in this figure; the direction of circumvention indicated in the figure is associated with the direction of the $x$-axis by the right-hand screw rule). Section 1 of the contour is opposite in direction to the $z$-axis. Therefore, $a_l$ on this section coincides with $-a_z$. Similar reasoning shows that $a_l$ on sections 2, 3, and 4 equals $a_y$, $a_z$ and $-a_y$, respectively. Hence, the circulation can be written in the form

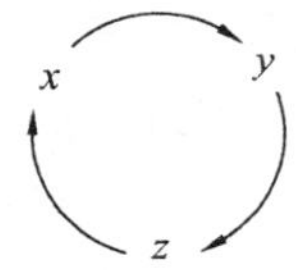


**Fig.8.28**

$$(a_{z,3} - a_{z,1})\Delta z - (a_{y,4} - a_{y,2})\Delta y \qquad (8.86)$$

where $a_{z,3}$ and $a_{z,1}$ are the average values of $a_z$ on sections 3 and 1, respectively, and $a_{y,4}$ and $a_{y,2}$ are the average values of $a_y$ on sections 4 and 2.

The difference $a_{z,3} - a_{z,1}$ is the increment of the average value of $a_z$ on the section $\Delta z$ when this section is displaced in the direction of the $y$-axis by $\Delta y$. Owing to the smallness of $\Delta y$ and $\Delta z$, this increment can be represented in the form $(\partial a_z/\partial y)\Delta y$, where the value of $\partial a_z/\partial y$ is taken for point $P$. Similarly, the difference $a_{y,4} - a_{y,2}$ can be represented in the form $(\partial a_y / \partial z)\Delta z$. Using these expressions in Eq. (8.86) and putting the common factor outside the parentheses, we get the following expression for the circulation:

$$\left(\frac{\partial a_z}{\partial y} - \frac{\partial a_y}{\partial z}\right)\Delta y \Delta z = \left(\frac{\partial a_z}{\partial y} - \frac{\partial a_y}{\partial z}\right)\Delta S$$

where $\Delta S$ is the area of the contour. Dividing the circulation by $\Delta S$, we find the expression for the projection of curl $\boldsymbol{a}$ onto the $x$-axis:

$$(\text{curl } \boldsymbol{a})_x = \frac{\partial a_z}{\partial y} - \frac{\partial a_y}{\partial z} \qquad (8.87)$$

We can find by similar reasoning that

$$(\text{curl } \boldsymbol{a})_y = \frac{\partial a_x}{\partial z} - \frac{\partial a_z}{\partial x} \qquad (8.88)$$

$$(\text{curl } \boldsymbol{a})_z = \frac{\partial a_y}{\partial x} - \frac{\partial a_x}{\partial y} \qquad (8.88)$$

It is easy to see that any of the equations (8.87) – (8.89) can be obtained from the preceding one [Eq. (8.89) should be considered as the preceding one for Eq. (8.87)] by the so-called cyclic transposition of the coordinates, i.e. by replacing the coordinates according to the scheme.

Thus, the curl of the vector $\boldsymbol{a}$ is determined in the Cartesian coordinate system by the following expression:

$$\text{curl}\boldsymbol{a} = \boldsymbol{e}_x\left(\frac{\partial a_z}{\partial y} - \frac{\partial a_y}{\partial z}\right) + \boldsymbol{e}_y\left(\frac{\partial a_x}{\partial z} - \frac{\partial a_z}{\partial x}\right) + \boldsymbol{e}_z\left(\frac{\partial a_y}{\partial x} - \frac{\partial a_x}{\partial y}\right) \qquad (8.90)$$

Below we shall indicate a more elegant way of writing this expression.

## Stokes's Theorem

Knowing the curl of the vector $\boldsymbol{a}$ at every point of surface $S$ (not necessarily plane), we can calculate the circulation of this vector around contour $\Gamma$ enclosing $S$ (the contour may also not be plane). For this purpose, we divide the surface into very small elements $\Delta S$. Owing to their smallness, these elements can be considered as plane. Therefore in accordance with Eq. (8.85), the circulation of the vector $\boldsymbol{a}$ around the contour bounding $\Delta S$ can be written in the form

$$\Delta C \approx (\text{curl}\boldsymbol{a})_n \Delta S = \text{curl}\boldsymbol{a} \cdot \Delta \boldsymbol{S} \tag{8.91}$$

where $\boldsymbol{n}$ is a positive normal to surface element $\Delta S$.

In accordance with Eq. (8.83), summation of expression (8.91) over all the $\Delta S$'s yields the circulation of the vector $\boldsymbol{a}$ around contour $\Gamma$ enclosing $S$:

$$C = \sum \Delta C \approx \sum \text{curl } \boldsymbol{a} \cdot \Delta \boldsymbol{S}$$

Performing a limit transition in which all the $\Delta S$'s shrink to zero (their number grows unlimitedly), we arrive at the equation

$$\oint_{\Gamma} \boldsymbol{a} \mathrm{d}\boldsymbol{l} = \int_{S} \text{curl } \boldsymbol{a} \cdot \mathrm{d}\boldsymbol{S} \tag{8.92}$$

Equation (8.92) is called Stokes's theorem. Its meaning is that the circulation of the vector $\boldsymbol{a}$ around an arbitrary contour $\Gamma$ equals the flux of the vector curl $\boldsymbol{a}$ through the arbitrary surface $S$ surrounded by the given contour.

## The Del Operator

Writing of the formulas of vector analysis is simplified quite considerably if we introduce a vector differential operator designated by the symbol $\nabla$ (nabla or del) and called the del operator or the Hamiltonian operator. This operator denotes a vector with the components $\partial/\partial x$, $\partial/\partial y$, and $\partial/\partial z$. Consequently,

$$\nabla = \boldsymbol{e}_x \frac{\partial}{\partial x} + \boldsymbol{e}_y \frac{\partial}{\partial y} + \boldsymbol{e}_z \frac{\partial}{\partial z} \tag{8.93}$$

This vector has no meaning by itself. It acquires a meaning, in combination with the scalar or vector function by which it is symbolically multiplied. Thus, if we multiply the vector $\nabla$ by the scalar $\varphi$, we obtain the vector

$$\Delta\varphi = \mathrm{e}_x \frac{\partial \varphi}{\partial x} + \mathrm{e}_y \frac{\partial \varphi}{\partial y} + \mathrm{e}_z \frac{\partial \varphi}{\partial z} \qquad (8.94)$$

which is the gradient of the function $\varphi$ [see Eq. (8.63)].

The scalar product of the vectors $\nabla$ and $a$ gives the scalar

$$\nabla \boldsymbol{a} = \nabla_x a_x + \nabla_y a_y + \nabla_z a_z = \frac{\partial a_x}{\partial x} + \frac{\partial a_x}{\partial x} + \frac{\partial}{\partial y} + \frac{\partial a_z}{\partial z} \qquad (8.95)$$

which we can see to be the divergence of the vector $\boldsymbol{a}$ [see Eq. (8.76)].

Finally, the vector product of the vectors $\nabla$ and $\boldsymbol{a}$ gives a vector with the components $[\nabla \boldsymbol{a}]_x = \nabla_y a_z - \nabla_z a_y = \partial a_z/\partial y - \partial a_y/\partial z$, etc., that coincide with the components of curl $\boldsymbol{a}$ [see Eqs. (8.87) – (8.89)]. Hence, using the writing of a vector product with the aid of a determinant, we have

$$\operatorname{curl}\boldsymbol{a} = [\nabla \boldsymbol{a}] = \begin{vmatrix} \boldsymbol{e}_x & \boldsymbol{e}_y & \boldsymbol{e}_z \\ \frac{\partial}{\partial x} & \frac{\partial}{\partial y} & \frac{\partial}{\partial z} \\ a_x & a_y & a_z \end{vmatrix} \qquad (8.96)$$

Thus, there are two ways of denoting the gradient, divergence, and curl:

$$\nabla\varphi \equiv \operatorname{grad}\varphi, \nabla\boldsymbol{a} \equiv \operatorname{div}\boldsymbol{a}, [\nabla\boldsymbol{a}] \equiv \operatorname{curl}\boldsymbol{a}$$

The use of the del symbol has a number of advantages. We shall therefore use such symbols in the following. One must accustom oneself to identify the symbol $\nabla\varphi$ with the words "gradient of phi" (i. e. to say not "del phi" but "gradient of phi"), the symbol $\nabla\boldsymbol{a}$ with the words "divergence of $\boldsymbol{a}$" and. finally, the symbol $\nabla\boldsymbol{a}$ with the words "curl of $\boldsymbol{a}$".

When using the vector $\nabla$, one must remember that it is a differential operator acting on all the functions to the right of it. Consequently, in transforming expressions including $\nabla$, one must take into consideration both the rules of vector algebra and those of differential calculus. For example, the derivative of the product of the functions $\varphi$ and $\psi$ is

$$(\varphi\psi)' = \varphi'\psi + \varphi\psi'$$

Accordingly,

$$\mathrm{grad}(\varphi\psi) = \nabla(\varphi\psi) = \psi\nabla\varphi + \varphi\nabla\psi = \psi\mathrm{grad}\varphi + \varphi\mathrm{grad}\psi \tag{8.97}$$

Similarly

$$\mathrm{div}(\varphi\boldsymbol{a}) = \nabla(\varphi\boldsymbol{a}) = \boldsymbol{a}\nabla\varphi + \varphi\nabla\boldsymbol{a} = \boldsymbol{a}\mathrm{grad}\varphi + \varphi\mathrm{div}\boldsymbol{a} \tag{8.98}$$

The gradient of a function $\varphi$ is a vector function. Therefore, the divergence and curl operations can be performed with it:

$$\mathrm{div\ grad}\ \varphi = \nabla(\nabla\varphi) = (\nabla\nabla)\varphi = (\nabla_x^2 + \nabla_y^2 + \nabla_z^2)\varphi = \frac{\partial^2}{\partial x^2} + \frac{\partial^2}{\partial y^2} + \frac{\partial^2}{\partial z^2} = \Delta\varphi \tag{8.99}$$

($\Delta$ is the Laplacian operator)

$$\mathrm{curl\ grad}\ \varphi = [\nabla, \nabla\varphi] = [\nabla\nabla]\varphi = 0 \tag{8.100}$$

(we remind our reader that the vector product of a vector and itself is zero).

Let us apply the divergence and curl operations to the function curl $\boldsymbol{a}$:

$$\mathrm{div\ curl}\ \boldsymbol{a} = \nabla[\nabla\boldsymbol{a}] = 0 \tag{8.101}$$

(a scalar triple product equals the volume of a parallelepiped constructed on the vectors being multiplied ; if two of these vectors coincide, the volume of the parallelepiped equals zero);

$$\mathrm{curl\ curl}\ \boldsymbol{a} = [\nabla, [\nabla\boldsymbol{a}]] = \nabla(\nabla\boldsymbol{a}) - (\nabla\nabla)\boldsymbol{a} = \mathrm{grad\ div}\ \boldsymbol{a} - \Delta\boldsymbol{a} \tag{8.102}$$

Equation (8.101) signifies that the field of a curl has no sources. Hence, the lines of the vector $[\nabla\ \boldsymbol{a}]$ have neither a beginning nor an end. It is exactly for this reason that the flux of a curl through any surface $S$ resting on the given contour $\Gamma$ is the same [see Eq. (8.92)].

We shall note in concluding that when the del operator is used, Eqs. (8.77) and (8.92) can be given the form

$$\oint_S \boldsymbol{a} \cdot \mathrm{d}\boldsymbol{S} = \int_V \nabla\boldsymbol{a} \cdot \mathrm{d}V \text{(the Ostrogradsky – Gauss theorem)} \tag{8.103}$$

$$\oint_{\Gamma} \boldsymbol{a} \cdot \mathrm{d}\boldsymbol{l} = \int_{S} [\nabla \boldsymbol{a}] \cdot \mathrm{d}\boldsymbol{S} \text{(stoke's theorem)} \qquad (8.104)$$

## Summary of Key Terms

**Electrostatics** The study of electric charges at rest relative to one another(not in motion, as in electric currents).

**Capacitor** An electrical device, in its simplest form a pair of parallel conducting plates separated by a small distance, that stores electric charge.

**Coulomb's law** The relationship among electrical force, charge, and distance $F = kq_1q_2/d^2$ If the charges are alike are in sign, the force is repelling; if the charges are unlike, the force is attractive.

**Coulomb** The SI unit of electrical charge. One coulomb is equal to the total charge of $6.25 \times 10^{18}$ electrons.

**Conductor** Any material that through which charge easily flows when subject to an external electrical force.

**Insulator** Any material that resists charge flow through it when subject to an external electrical force.

**Semiconductor** A poorly conducting material, such as crystalline silicon or germanium, that can be made a better-conducting material by the addition of certain impurities or energy.

**Charging by contact** The transfer of charge from one substance to another by physical contact between substances.

**Charging by induction** The change in charge of a grounded object, caused by the electrical influence of electric charge close by but not in contact.

**Electrically polarized** Term applied to an atom or molecule in which the charges are aligned so that one side is slightly more positive or negative than the opposite side.

**Electric field** The energetic region of space surrounding a charged object. About a charged point, the field decreases with distance according to the inverse square law, like a gravitational field. Between oppositely charged parallel plates, the electric field is uniform. A charged object placed in the region of an electric field experiences a force.

**Electric potential energy** The energy a charge possesses by virtue of its location in an electric field.

**Electric potential** The electric potential energy per amount of charge, measured in volts, and often called voltage:

$$\text{Voltage} = \frac{\text{electric energy}}{\text{amount of charge}}$$

# 9

# Magnetic Field in a Vacuum

## 9.1 INTERACTION OF CURRENTS

Experiments show that electric currents exert a force on one another. For example, two thin straight parallel conductors carrying a current (we shall call them line currents) attract each other if the currents in them flow in same directions, and repel each other if the currents flow in opposite directions. The force of interaction per unit length of each of the parallel conductors is proportional to the magnitudes of the currents $I_1$ and $I_2$ in them and inversely proportional to the distance $b$ between them:

$$F_u = k \frac{2I_1 I_2}{b} \tag{9.1}$$

We have designated the proportionality constant $2k$ for reasons that will become clear on a later page.

The law of interaction of currents was established in 1820 by the French physicist Andre Ampere (1775 ~ 1836). A general expression of this law suitable for conductors of any shape will be given in Sec.9.6.

Equation (9.1) is used to establish the unit of current in the SI and in the absolute electromagnetic system (cgsm) of units. The SI unit of current the ampere is defined as the constant current which, if maintained in two straight parallel conductors of infinite length, of negligible cross section, and placed 1 meter apart in vacuum, would produce between these conductors a force equal to $2 \times 10^{-7}$ Newton per meter of length.

The unit of charge, called the coulomb, is defined as the charge passing in 1 second through the cross section of a conductor in which a constant current of 1 ampere is flowing. Accordingly, the coulomb is also called the ampere – second (A·s).

Equation (9.1) is written in the rationalized form as follows:

$$F_u = \frac{\mu_0}{4\pi} \frac{2I_1 I_2}{b} \tag{9.2}$$

where $\mu_0$ is the so – called magnetic constant [compare with Eq. (8.11)]. To find the numerical value of $\mu_0$ we shall take advantage of the fact that according to the definition of the ampere, when $I_1 = I_2 = 1$ A and $b = 1$ m, the force Fu is obtained equal to $2 \times 10^{-7}$ N/m. Let us use these values in Eq. (9.2):

$$2 \times 10^{-7} = \frac{\mu_0}{4\pi} \frac{2 \times 1 \times 1}{1}$$

Hence,

$$\mu_0 = 4\pi \times 10^{-7} = 1.26 \times 10^{-6} \text{ H/m} \tag{9.3}$$

The constant $k$ in Eq. (9.1) can be made equal to unity by choosing an appropriate unit of current. This is how the absolute electromagnetic unit of current ($\text{cgsm}_I$) is established. It is defined as the current which, if maintained in a thin straight conductor of infinite length, would act on an equal and parallel line current at a distance of 1 cm from it with a force equal to 2 dyn per centimeter of length .

In the cgse system, the constant $k$ is a dimension quantity other than unity. According to Eq. (9.1), the dimension of $k$ is determined as follows:

$$[k] = \frac{[F_u b]}{[I]^2} = \frac{[F]}{[I]^2} \tag{9.4}$$

We have taken into account that the dimension of $F_u$ is the dimension of force divided by the dimension of length; hence, the dimension of the product $E_u b$ is that of force. According to Eq. (8.7)

$$[F] = \frac{[q]^2}{L^2}; \quad [I] = \frac{[q]}{T}$$

Using these values in Eq. (9.4), we find that

$$[k] = \frac{T^2}{L^2}$$

Consequently, in the cgse system, $k$ can be written in the form

$$k = \frac{1}{c^2} \tag{9.5}$$

where $c$ is a quantity having the dimension of velocity and called the electromagnetic constant. To find its value, let us use relation (8.8) between the coulomb and the cgse unit of charge, which was established experimentally. A force of $2 \times 10^{-7}$ N/m is equivalent to $2 \times 10^{-4}$ dyn/cm. According to Eq. (9.1), this is the force with which currents of $3 \times 10^9$ $\text{cgse}_I$(i.e. 1 A) each interact when $b = 100$ cm. Thus,

$$2 \times 10^{-4} = \frac{1}{c^2}\frac{2 \times 3 \times 10^9 \times 3 \times 10^9}{100}$$

Whence

$$c = 3 \times 10^{10} \text{ cm/s} = 3 \times 10^8 \text{ m/s} \tag{9.6}$$

The value of the electromagnetic constant coincides with that of the speed of light in a vacuum. From J. Maxwell's theory, there follows the existence of electromagnetic waves whose speed in a vacuum equals the electromagnetic constant $c$. The coincidence of $c$ with the speed of light in a vacuum gave Maxwell the grounds to assume that light is an electromagnetic wave.

The value of $k$ in Eq. (9.1) is 1 in the cgsm system and $1/c^2 = 1/$

$(3 \times 10^{10})^2$ $s^2/cm^2$ in the cgsm system. Hence it follows that a current of 1cgsm is equivalent to a current of $3 \times 10^{10}$ $cgse_I$

$$1\ cgsm_I = 3 \times 10^{10}\ cgse_I = 10\ A \tag{9.7}$$

Multiplying this relation by 1 s, we get

$$1\ cgsm_q = 3 \times 10^{10}\ cgse_q = 10\ C \tag{9.8}$$

Thus,

$$I_{cgsm} = \frac{1}{c} I_{cgse} \tag{9.9}$$

Accordingly,

$$q_{cgsm} = \frac{1}{c} q_{cgse} \tag{9.10}$$

There is a definite relation between the constants $\varepsilon_0$, $\mu_0$ and $c$. To establish it, let us find the dimension and numerical value of the product $\varepsilon_0\mu_0$. In accordance with Eq. (8.11), the dimension of $\varepsilon_0$ is

$$[\varepsilon_0] = \frac{[q]^2}{L^2[F]} \tag{9.11}$$

According to Eq. (9.2)

$$[\mu_0] = \frac{[F_u b]}{[I]^2} = \frac{[F]T^2}{[q]^2} \tag{9.12}$$

Multiplication of Eqs. (9.11) and (9.12) yields

$$[\varepsilon_0\mu_0] = \frac{T^2}{L^2} = \frac{1}{[v]^2} \tag{9.13}$$

($v$ is the speed).

With account taken of Eqs. (8.12) and (9.3), the numerical value of the product $\varepsilon_0\mu_0$ is

$$\varepsilon_0\mu_0 = \frac{1}{4\pi \times 9 \times 10^9} \times 4\pi \times 10^{-7} = \frac{1}{(3 \times 10^8)^2} \frac{s^2}{m^2} \tag{9.14}$$

Finally, taking into account Eqs. (9.6), (9.13), and (9.14), we get the relation interesting us:

$$\varepsilon_0\mu_0 = \frac{1}{c^2} \tag{9.15}$$

## 9.2 MAGNETIC FIELD

Currents interact through a field called magnetic. This name originated from the fact that, as the Danish physicist Hans Oersted (1777 ~ 1851) discovered in 1820, the field set up by a current has an orienting action on a magnetic pointer. Oersted stretched a wire carrying a current over a magnetic pointer rotating on a needle. When the current was switched on, the pointer aligned itself at right angles to the wire. Reversing of the current caused the pointer to rotate in the opposite direction.

Oersted's experiment shows that a magnetic field has a sense of direction and must be characterized by a vector quantity. The latter is designated by the symbol $\boldsymbol{B}$. It would be logical to call $\boldsymbol{B}$ the magnetic field strength, by analogy with the electric field strength $\boldsymbol{E}$. For historical reasons, however, the basic force characteristic of a magnetic field was called the magnetic induction. The name magnetic field strength was given to an auxiliary quantity $\boldsymbol{H}$ similar to the auxiliary characteristic $\boldsymbol{D}$ of an electric field.

A magnetic field, unlike its electric counterpart, does not act on a charge at rest. A force appears only when a charge is moving.

A current-carrying conductor is an electrically neutral system of charges in which the charges of one sign are moving in one direction, and the charges of the other sign in the opposite direction (or are at rest). It thus follows that a magnetic field is set up by moving charges.

Thus, moving charges (currents) change the properties of the space surrounding them—they set up a magnetic field in it. This field manifests itself in that forces are exerted on charges moving in it (currents).

Experiments show that the superposition principle holds for a magnetic field, the same as for an electric field: the field $\boldsymbol{B}$ set up by several moving charges (currents) equals the vector sum of the fields $\boldsymbol{B}_i$ set up by each charge (current) separately:

$$\boldsymbol{B} = \sum \boldsymbol{B}_i \tag{9.16}$$

compare with Eq. (8.19).

## 9.3 FIELD OF A MOVING CHARGE

Space is isotropic, consequently, if a charge is stationary, then all directions have equal rights. This underlies the fact that the electrostatic field set up by a point charge is spherically symmetrical.

If a charge travels with the velocity $\boldsymbol{v}$, a preferred direction (that of the vector $\boldsymbol{v}$) appears in space. We can therefore expert the magnetic field produced by a moving charge to have axial symmetry. We must note that we have in mind free motion of a charge, i.e. motion with a constant velocity. For an acceleration to appear, the charge must experience the action of a field (electric or magnetic). This field by its very existence would violate the isotropy of space.

Let us consider the magnetic field set up at point $P$ by the point charge $q$ travelling with the constant velocity $\boldsymbol{v}$ (Fig. 9.1) The disturbances of the field are transmitted from point to point with the finite velocity $c$. For this reason, the induction $\boldsymbol{B}$ at point $P$ at the moment of time $t$ is determined not by the position of the charge at the same moment $t$, but by its position at an earlier moment of time $t - \tau$:

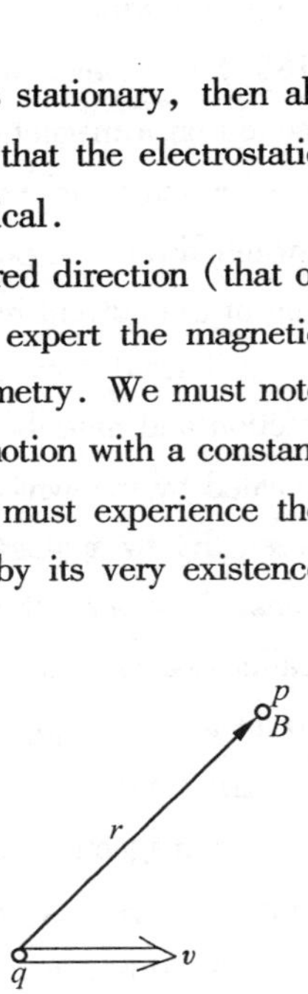

Fig.9.1

$$\boldsymbol{B}(P, t) = f\{q, \boldsymbol{v}, r(t - \tau)\}$$

Here $P$ signifies the collection of the coordinates of point $\boldsymbol{P}$ determined in a stationary reference frame, and $\boldsymbol{r}(t - \tau)$ is the position vector drawn to point $P$ from the point where the charge was at the moment $t - \tau$.

If the velocity of the charge is much smaller than $c$ ($v \ll c$), then the retardation time $\tau$ will be negligibly small. In this case, we can consider that the value of $\boldsymbol{B}$ at the moment $t$ is determined by the position of the charge at the same moment $t$. If this condition is observed, then

$$\boldsymbol{B}(P,t) = f\{q,\boldsymbol{v},\boldsymbol{r}(t)\} \tag{9.17}$$

The form of function (9.17) can be established only experimentally. But before giving the results of experiments. let us try to find the logical form of this relation. The simplest, assumption is that the magnitude of the vector $\boldsymbol{B}$ is proportional to the charge $q$ and the velocity $\boldsymbol{v}$ (when $\boldsymbol{v} = 0$, a magnetic field is absent). We have to "construct" the vector $\boldsymbol{B}$ we are interested in from the scalar $q$ and the two given vectors $\boldsymbol{v}$ and $\boldsymbol{r}$. This can be done by vector multiplication of the given vectors and then by multiplying their product by the scalar. The result is the express

$$q[\boldsymbol{vr}] \tag{9.18}$$

The magnitude of this expression grows with an increasing distance from the charge (with increasing $r$). It is improbable that the characteristic of a field will behave in this way — for the fields that we know (electrostatic, gravitational), the field does not grow with an increasing distance from the source, but, on the contrary, weakens, varying in proportion to $1/r^2$. Let us assume that the magnetic field of a moving charge behaves in the same way when r changes. We can obtain an inverse proportion to the square of $r$ by dividing Eq. (9.18) by $r^3$. The result is

$$\frac{q[\boldsymbol{vr}]}{r^3} \tag{9.19}$$

Experiments show that when $v \ll c$, the magnetic induction of the field of a moving charge is determined by the formula

$$\boldsymbol{B} = k' \frac{q[\boldsymbol{vr}]}{r^3} \tag{9.20}$$

Where $k'$ is a proportionality constant.

We must stress once more that the reasoning which led us to expression (9.19) must by no means be considered as the derivation of Eq. (9.20). This reasoning does not have conclusive force. Its aim is to help us understand and memorize Eq. (9.20). This equation itself can be obtained only experimentally.

It can be seen from Eq. (9.20) that the vector $\boldsymbol{B}$ at every point $P$ is directed at right angles to the plane passing through the direction of the vector $\boldsymbol{v}$ and point $P$, so that rotation in the direction of $\boldsymbol{B}$ forms a right-handed systems with the direction of $\boldsymbol{v}$ (see the circle with the dot in Fig. 9.1). We must note that $\boldsymbol{B}$ is a pseudo vector.

The value of the proportionality constant $k'$ depends on our choice of the units of the quantities in Eq. (9.20). This equation is written in the rationalized form as follows:

$$\boldsymbol{B} = \frac{\mu_0}{4\pi}\frac{q[\boldsymbol{vr}]}{r^3} \tag{9.21}$$

This equation can be written in the form

$$\boldsymbol{B} = \frac{\mu_0}{4\pi}\frac{q[\boldsymbol{ve}_r]}{r^2} \tag{9.22}$$

[compare with Eq. (8.15)]. It must be noted that in similar equations when $\varepsilon_0$ is in the denominator, $\mu_0$ is in the numerator, and vice versa.

The SI unit of magnetic induction is called the tesla ($T$) in honour of the Croatian electrician and inventor Nikola Tesia (1856 ~ 1943).

The units of the magnetic induction $B$ are chosen in the cgse and cgsm systems so that the constant $k'$ in Eq. (9.20) equals unity. Hence, the same relation holds between the units of $B$ in these systems as between the units of charge:

$$1\ \text{cgsm}_B = 3 \times 10^{10}\ \text{cgse}_B \tag{9.23}$$

[see Eq. (9.8)].

The cgsm unit of magnetic induction has a special name — the gauss (Gs).

The German mathematician Karl Gauss (1777 ~ 1855) proposed a system of units in which all the electrical quantities (charge, current, electric field strength, etc.) are measured in cgse units, and all the magnetic quantities (magnetic induction, magnetic moment, etc.) in cgsm units. This system of units was named the Gaussian one, in honor of its author.

In the Gaussian system, owing to Eqs. (9.9) and (9.10), all the equations containing the current or charge in addition to magnetic quantities include one multiplier $1/c$ for each quantity $I$ or $q$ in the relevant equation. This multiplier converts the value of the pertinent quantity ($I$ or $q$) expressed in cgse units to a value expressed in cgsm units (the cgsm system of units is constructed so that the proportionality constants in all the equations equal 1). For example, in the Gaussian system, Eq. (9.20) has the form

$$\boldsymbol{B} = \frac{1}{c}\frac{q[\boldsymbol{vr}]}{r^3} \tag{9.24}$$

We must note that the appearance of a preferred direction in space (the direction of the vector $\boldsymbol{v}$) when a charge moves leads to the electric field of the moving charge also losing its spherical symmetry and becoming axially symmetrical. The relevant calculations show that the $\boldsymbol{E}$ lines of the field of a freely moving charge have the form shown in Fig. 9.2. The vector $\boldsymbol{E}$ at point $P$ is directed along the position vector $\boldsymbol{r}$ drawn from the point where the charge is at the given moment to point $P$. The magnitude of the field strength is determined by the equation

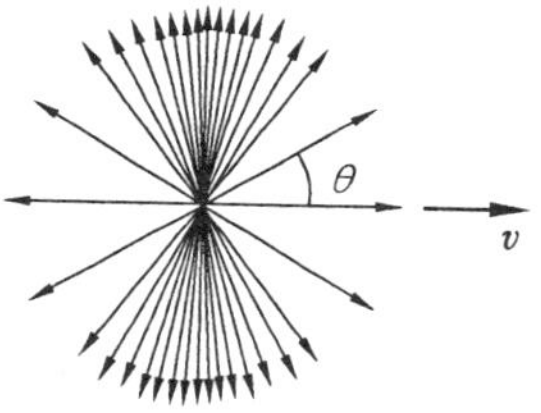

**Fig.9.2**

$$E = \frac{1}{4\pi\varepsilon_0}\frac{q}{r^2}\frac{1 - v^2/c^2}{[1 - (v^2/c^2)\sin^2\theta]^{3/2}} \tag{9.25}$$

where $\theta$ is the angle between the direction of the velocity $\boldsymbol{v}$ and the position vector $\boldsymbol{r}$.

When $v \ll c$, the electric field of a freely moving charge at each moment of time dons not virtually differ from the electrostatic field set up by a stationary charge at the point where the moving charge is at the given moment. It must be remembered, however that this "electrostatic" field moves together with the charge. Hence the field at each point of space changes with time.

At values of $v$ comparable with $c$, the field in directions at right angles to $\boldsymbol{v}$ is appreciably stronger than in the direction of motion at the same distance from the charge (see Fig.9.2 drawn for $v/c = 0.8$) The field "flattens out" in the direction of motion and is concentrated mainly near a plane passing through the charge and perpendicular to the vector $\boldsymbol{v}$.

## 9.4 THE BIOT-SAVART LAW

Let us determine the nature of the magnetic field set up by an arbitrary thin wire through which a current flows. We shall consider a small element of the wire of length $\mathrm{d}l$. This element contains $nS\mathrm{d}l$ current carriers ($n$ is the number of carriers in a unit volume, and $S$ is the cross-sectional area of the wire where the element $\mathrm{d}l$ has been taken). At the point whose position relative to the element $\mathrm{d}l$ is determined by the position vector $\boldsymbol{r}$ (Fig.9.3), a separate carrier of current $e$ sets up a field with the induction

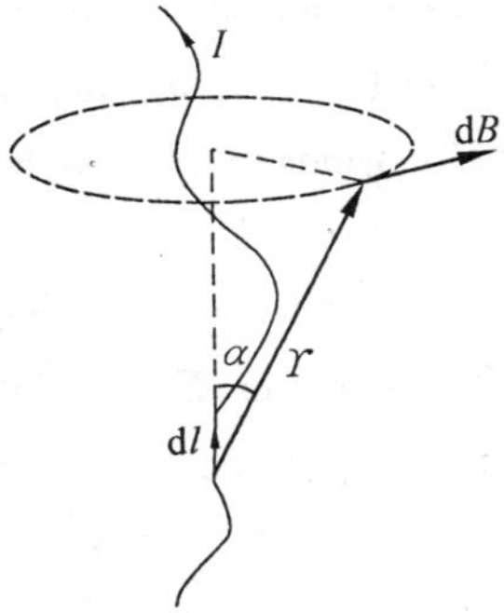


**Fig.9.3**

$$\boldsymbol{B} = \frac{\mu_0}{4\pi}\frac{e[(\boldsymbol{v}+\boldsymbol{u}),\boldsymbol{r}]}{r^3}$$

[see Eq.(9.21)]. Here $\boldsymbol{v}$ is the velocity of chaotic motion, and $\boldsymbol{u}$ is the velocity of ordered motion of the carrier.

The value of the magnetic induction averaged over the current carriers in the element $\mathrm{d}l$ is

$$<\boldsymbol{B}> = \frac{\mu_0}{4\pi}\frac{e[(<\boldsymbol{v}>+<\boldsymbol{u}>),r]}{r^3} = \frac{\mu_0}{4\pi}\frac{e[<\boldsymbol{u}>,\boldsymbol{r}]}{r^3}$$

$(<\boldsymbol{v}> = 0)$. Multiplying this expression by the number of carriers in an element of the wire (equal to $nS\mathrm{d}l$), we get the contribution to the field introduced by the element $\mathrm{d}\boldsymbol{l}$:

$$\mathrm{d}\boldsymbol{B} = \langle \boldsymbol{B} \rangle nS\mathrm{d}l = \frac{\mu_0}{4\pi}\frac{S[(ne\langle \boldsymbol{u} \rangle), \boldsymbol{r}]\mathrm{d}l}{r^3}$$

(we have put the scalar multipliers n and e inside the sign of the vector product). Taking into account that ne $\langle \boldsymbol{u} \rangle = \boldsymbol{j}$, we can write

$$\mathrm{d}\boldsymbol{B} = \frac{\mu_0}{4\pi}\frac{S[\boldsymbol{j}, \boldsymbol{r}]\mathrm{d}l}{r^3} \tag{9.26}$$

Let us introduce the vector $\mathrm{d}\boldsymbol{l}$ directed along the axis of the current element $\mathrm{d}l$ in the same direction as the current. The magnitude of this vector is $\mathrm{d}l$. Since the directions of the vectors $\boldsymbol{j}$ and $\mathrm{d}\boldsymbol{l}$ coincide, we can write the equation

$$\boldsymbol{j}\mathrm{d}l = j\mathrm{d}\boldsymbol{l} \tag{9.27}$$

Performing such a substitution in Eq. (9.26), we get

$$\mathrm{d}\boldsymbol{B} = \frac{\mu_0}{4\pi}\frac{S_j[\mathrm{d}\boldsymbol{l}, \boldsymbol{r}]}{r^3}$$

Finally, taking into account that the product $S_j$ gives the current $I$ In the wire, we arrive at the final expression determining the magnetic induction of the field set up by a current element of length $\mathrm{d}\boldsymbol{l}$:

$$\mathrm{d}\boldsymbol{B} = \frac{\mu_0}{4\pi}\frac{I[\mathrm{d}\boldsymbol{l}, \boldsymbol{r}]}{r^3} \tag{9.28}$$

We have derived Eq.(9.28) from Eq.(9.21). Equation (9.28) was actually established experimentally before Eq.(9.21) was known. Moreover, the latter equation was derived from Eq.(9.28).

In 1820, the French physicists Jean Biot (1774 ~ 1862) and Felix Savart (1791 ~ 1841) studied the magnetic fields flowing along thin wires of various shape. The French astronomer and mathematician Pierre La place (1749 ~ 1827) analyzed the experimental data obtained and found that the magnetic held of any current can be calculated as the vector sum (superposition) of the fields set up by the separate elementary sections of the currents. Laplace obtained Eq. (9.28) for the magnetic induction of the field set up by a current element of length $d\boldsymbol{l}$. In this connection, Eq. (9.28) is called the **BiotSavart – Laplace Law**, or more briefly the **Biot – Savart Law**.

A glance at Fig. 9.3 shows that the vector $d\boldsymbol{B}$ is directed at right angled to the plane passing through $d\boldsymbol{l}$ and the point for which the field is being calculated so that rotation about $d\boldsymbol{l}$ in the direction of $d\boldsymbol{B}$ is associated with $d\boldsymbol{l}$ in the direction of $d\boldsymbol{B}$ is associated with $d\boldsymbol{l}$ by the right – hand screw rule. The magnitude of $d\boldsymbol{B}$ is determined by the expression

$$dB = \frac{\mu_0}{4\pi}\frac{Id\boldsymbol{l}\sin\alpha}{\boldsymbol{r}^2} \qquad (9.29)$$

where $\alpha$ is the angle between the vectors $d\boldsymbol{l}$ and $\boldsymbol{r}$.

Let us use Eq. (9.28) to calculate the field of a line current, i.e. the field set up by a current flowing through a thin straight wire of infinite length (Fig. 9.4). All the vectors $d\boldsymbol{B}$ at a given point have the same direction (in our case beyond the drawing). Therefore, addition of the vectors $d\boldsymbol{B}$ may be replaced with addition of their magnitudes. The point for which we are calculating the magnetic induction is at the distance $b$ from the wire.

Inspection of Fig. 9.4 shows that

$$r = \frac{b}{\sin\alpha}, \quad dl = \frac{r d\alpha}{\sin\alpha} = \frac{b d\alpha}{\sin^2\alpha}$$

Let us introduce these values into Eq. (9.29):

$$d\boldsymbol{B} = \frac{\mu_0}{4\pi}\frac{Ib d\alpha \sin\alpha \sin^2\alpha}{b^2 \sin^2\alpha} = \frac{\mu_0}{4\pi}\frac{I}{b}\sin\alpha\, d\alpha$$

The angle $\alpha$ varies within the limits from 0 to $\pi$ for all the elements of an infinite line current. Hence,

$$B = \int dB = \frac{\mu_0}{4\pi}\frac{I}{b}\int_0^{\pi}\sin\alpha\, d\alpha = \frac{\mu_0}{4\pi}\frac{2I}{b}$$

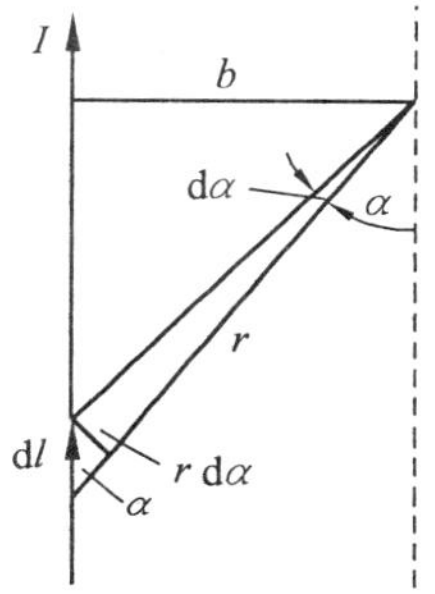


**Fig.9.4**

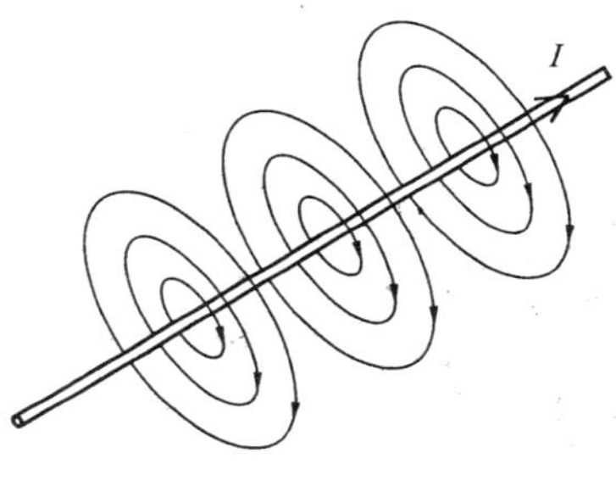


**Fig.9.5**

Thus, the magnetic induction of the field of a line current is determined by the formula

$$B = \frac{\mu_0}{4\pi}\frac{2I}{b} \qquad (9.30)$$

The magnetic induction lines of the field of a line current are a system of concentric circles surrounding the wire (Fig. 9.5).

## 9.5 THE LORENTZ FORCE

A charge moving in a magnetic field experiences a force which we shall call **magnetic**. The force is determined by the charge $q$, its velocity $\boldsymbol{v}$, and the magnetic induction $\boldsymbol{B}$ at the point where the charge is at the moment of time being considered. The simplest assumption is that the magnitude of the force $F$ is proportional to each of the three quantities $q$, $v$, and $B$. In addition, $F$ can be expected to depend on the mutual orientation of the vectors $\boldsymbol{v}$ and $\boldsymbol{B}$. The direction of the vector $\boldsymbol{F}$ should be determined by those of the vectors $\boldsymbol{v}$ and $\boldsymbol{B}$.

To "construct" the vector $\boldsymbol{F}$ from the scalar $q$ and the vectors $\boldsymbol{v}$ and $\boldsymbol{B}$, let us find the vector product of $\boldsymbol{v}$ and $\boldsymbol{B}$ and then multiply the result obtained by the scalar $q$. The result is the expression

$$q[\boldsymbol{vB}] \qquad (9.31)$$

It has been established experimentally that the force $\boldsymbol{F}$ acting on a charge moving in a magnetic field is determined by the formula

$$F = kq[\boldsymbol{vB}] \tag{9.32}$$

where $k$ is a proportionality constant depending on the choice of the units for the quantities in the formula.

It must be borne in mind that the reasoning which led us to expression(9.31) must by no means be considered as the derivation of Eq. (9.32). This reasoning does not have conclusive force. Its aim is to help us memorize Eq.(9.32). The correctness of this equation can be established only experimentally.

We must note that Eq. (9.32) can be considered as a definition of The magnetic induction $\boldsymbol{B}$.

The unit of magnetic induction $B$ – the tesla – is determined so that the proportionality constant $k$ in Eq.(9.32) equals unity. Hence, In SI units, this equation becomes

$$\boldsymbol{F} = q[\boldsymbol{vB}] \tag{9.33}$$

The magnitude of the magnetic force is

$$F = qvB\sin\alpha \tag{9.34}$$

Where $\alpha$ is the angle between the vectors $\boldsymbol{v}$ and $\boldsymbol{B}$. It can be seen from Eq.(9.34) that a charge moving along the lines of a magnetic field does not experience the action of a magnetic force.

The magnetic force is directed at right angles to the plane containing the vectors $\boldsymbol{v}$ and $\boldsymbol{B}$. If the charge $q$ is positive, then the direction of the force coincides with that of the vector $[\boldsymbol{vB}]$. When $q$ is negative, the directions of the vectors $\boldsymbol{F}$ and $[\boldsymbol{vB}]$ are opposite (Fig.9.6).

Since the magnetic force is always directed at right angles to the velocity of a charged particle, it does no work on the particle. Hence, we cannot change the energy of a charged particle by acting on it with a constant magnetic field.

The force exerted on a charged particle that is simultaneously in an electric and a magnetic field is

$$\boldsymbol{F} = q\boldsymbol{E} + q[\boldsymbol{vB}] \tag{9.35}$$

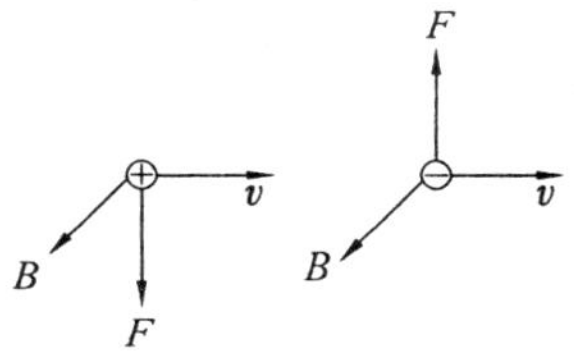


**Fig.9.6**

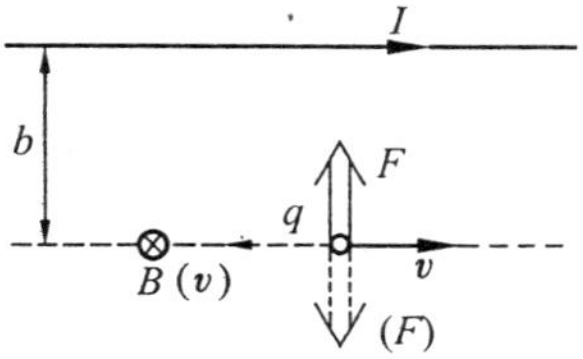


**Fig.9.7**

This expression was obtained from the results of experiments by the Dutch physicist Hendrik Lorentz (1853 ~ 1928) and is called the Lorentz force.

Assume that the charge $q$ is moving with the velocity $\boldsymbol{v}$ parallel to a straight infinite wire along which the current $I$ flows (Fig.9.7). According to Eqs. (9.30) and (9.34), the charge in this case experiences a magnetic force whose magnitude is

$$F = qvB = qv\frac{\mu_0}{4\pi}\frac{2I}{b} \tag{9.36}$$

where $b$ is the distance from the charge to the wire. The force is directed toward the wire when the charge is positive if the directions of the current and motion of the charge are the same, and away from the wire if these directions are opposite (see Fig.9.7). When the charge is negative, the direction of the force is reversed, the other conditions being equal.

Let us consider two like point charges $q_1$ and $q_2$ moving along parallel straight lines with the same velocity $v$ that is much smaller than $c$ (Fig.9.8). When $v \ll c$, the electric field does not virtually differ from the field of stationary charges (see Sec.9.3). Therefore the magnitude of the electric force $F_e$ exerted on the charges can be considered equal to

$$F_{e,1} = F_{e,2} = F_e = \frac{1}{4\pi\varepsilon_0}\frac{q_1q_2}{r^2} \tag{9.37}$$

Equations (9.21) and (9.33) give us the following expression for the magnetic force $F_m$ exerted on the charges:

$$F_{m,1} = F_{m,2} = F_m = \frac{\mu_0}{4\pi}\frac{q_1 q_2 v^2}{r^2} \tag{9.38}$$

(the position vector $\boldsymbol{r}$ is perpendicular to $\boldsymbol{v}$).

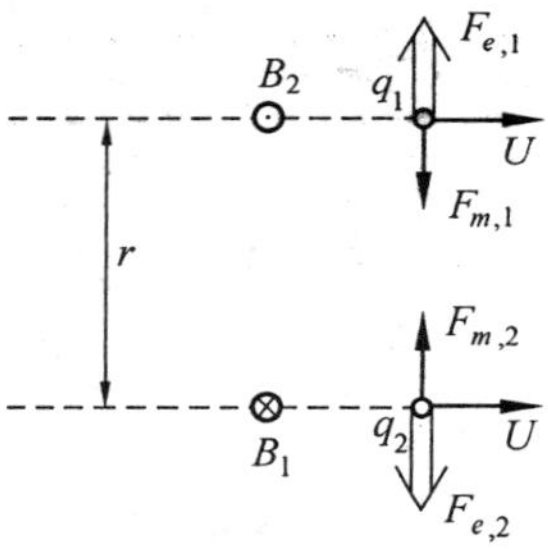


Fig.9.8

Let us find the ratio between the magnetic and electric forces It follows from Eqs. (9.37) and (9.38) that

$$\frac{F_m}{F_e} = \varepsilon_0\mu_0 v^2 = \frac{v^2}{c^2} \tag{9.39}$$

[see Eq (9.15)]. We have obtained Eq. (9.39) on the assumption that $v \ll c$. This ratio holds, however, with any $v$'s.

The forces $F_e$ and $F_m$ are directed oppositely. Figure 9.8 has been drawn for like and positive charges. For like negative charges, the directions of the forces will remain the same, while the directions of the vectors $\boldsymbol{B}_1$ and $\boldsymbol{B}_2$ will be reversed. For unlike charges, the directions of the electric and magnetic forces will be the reverse of those shown in the figure.

Inspection of Eq. (9.39) shows that the magnetic force is weaker than the Coulomb one by a factor equal to the square of the ratio of the speed of the charge to that of light. The explanation is that the magnetic interaction between moving charges is a relativistic effect. Magnetism would disappear if the speed of light were infinitely great.

## 9.6 AMPERE'S LAW

If a wire carrying a current is in a magnetic field, then each of the current carriers experiences the force

$$\boldsymbol{F} = e[(\boldsymbol{v} + \boldsymbol{u}), \boldsymbol{B}] \tag{9.40}$$

[see Eq. (9.33)]. Here $\boldsymbol{v}$ is the velocity of chaotic motion of a carrier and $u$ is the velocity of ordered motion. The action of this force is transferred from a current carrier to the conductor along which it is moving. As a result, a force acts on a wire with current in a magnetic field.

Let us find the value of the force $\mathrm{d}\boldsymbol{F}$ exerted on an element of a wire of length $\mathrm{d}l$. We shall average Eq. (9.40) over the current carriers contained in the element $\mathrm{d}l$:

$$\langle \boldsymbol{F} \rangle = e[(\langle \boldsymbol{v} \rangle + \langle \boldsymbol{u} \rangle), \boldsymbol{B}] = e[\langle \boldsymbol{u} \rangle, \boldsymbol{B}] \tag{9.41}$$

($\boldsymbol{B}$ is the magnetic induction at the place where the element $\mathrm{d}l$ is). The wire element contains $nS\mathrm{d}l$ carriers ($n$ is the number of carriers in unit volume, and $S$ is the cross-sectional area of the wire at the given place). Multiplying Eq. (9.41) by the number of carriers, we find the force we are interested in:

$$\mathrm{d}\boldsymbol{F} = \langle \boldsymbol{F} \rangle nS\mathrm{d}l = [(ne \langle \boldsymbol{u} \rangle), \boldsymbol{B}]S\mathrm{d}l$$

Taking into account that ne $\langle \boldsymbol{u} \rangle$ is the current density $\boldsymbol{j}$, and $S\mathrm{d}l$ gives the volume of a wire element $\mathrm{d}V$, we can write

$$\mathrm{d}\boldsymbol{F} = [\boldsymbol{jB}]\mathrm{d}V \tag{9.42}$$

Hence, we can obtain an expression for the density of the force, i.e. for the force acting on unit volume of the conductor

$$\boldsymbol{F}_{u,v} = [\boldsymbol{jB}] \tag{9.43}$$

Let us write Eq. (9.42) in the form

$$\mathrm{d}\boldsymbol{F} = [\boldsymbol{jB}]S\mathrm{d}l$$

Replacing in accordance with Eq. (9.27) $jS$ $\mathrm{d}l$ with $jS$ $\mathrm{d}\boldsymbol{l} = I\mathrm{d}\boldsymbol{l}$, we arrive at the equation

$$\mathrm{d}\boldsymbol{F} = I[\mathrm{d}\boldsymbol{l}, \boldsymbol{B}] \tag{9.44}$$

This equation determines the force exerted on a current element $\mathrm{d}\boldsymbol{l}$ in a magnetic field. Equation (9.44) was established experimentally by Ampere and is called Ampere's law.

We have obtained Ampere's law on the basis of Eq. (9.33) for the magnetic force. The expression for the magnetic force was actually obtained from the experimentally established equation (9.44).

The magnitude of the force (9.44) is calculated by the equation

$$\mathrm{d}F = IB\mathrm{d}l\sin\alpha \qquad (9.45)$$

where $\alpha$ is the angle between the vectors $\mathrm{d}\boldsymbol{l}$ and $\boldsymbol{B}$ (Fig.9.9). The forces normal to the plane containing the vectors $\mathrm{d}\boldsymbol{l}$ and $\boldsymbol{B}$.

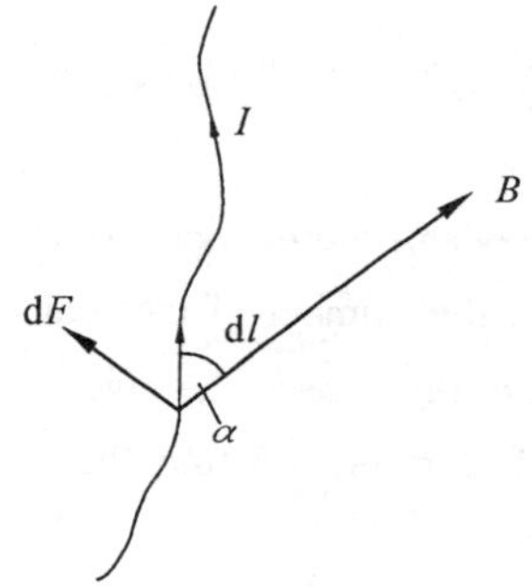


Fig.9.9

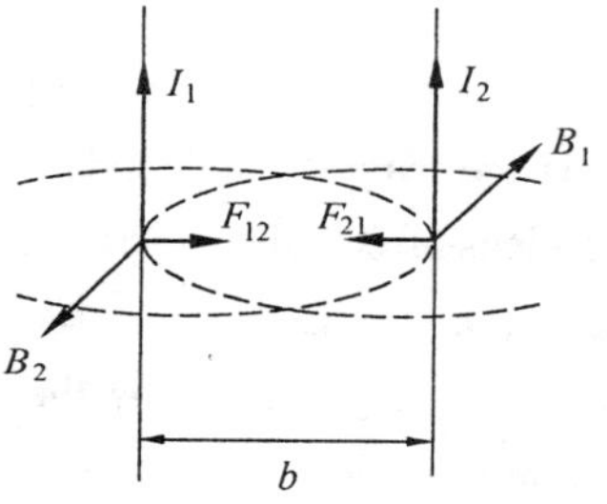


Fig.9.10

Let us use Ampere's law to calculate the force of interaction between two parallel infinitely long line currents in a vacuum. If the distance between the currents is $b$ (Fig.9.10), then each element of the current $I_2$ will be in a magnetic field whose induction is $B_1 = (\mu_0/4\pi)(2I_1/b)$ [see Eq.(9.30)]. The angle $\alpha$ between the elements of the current $I_2$ and the vector $B_1$ is a right one. Hence, according to Eq. (9.45), the force acting on unit length of the current $I_2$ is

$$F_{21,u} = I_2B_1 = B = \frac{\mu_0}{4\pi}\frac{2I_1I_2}{b} \qquad (9.46)$$

Equation (9.46) coincides with Eq.(9.2).

We get a similar equation for the force $F_{12,u}$ exerted on unit length of the current $I_1$. It is easy to see that when the currents flow in the same direction they attract each other, and in the opposite direction repel each other.

## The Further Study:

### Magnetism as a Relativistic Effect

There is a deep relation between electricity and magnetism. On the basis of the postulates of the theory of relativity and of the invariance of an electric charge, we can show that the magnetic interaction of charges and currents is a corollary of Coulomb's law. We shall show this on the example of a charge moving parallel to an infinite-line current with the velocity $v_0$(Fig.9.11).

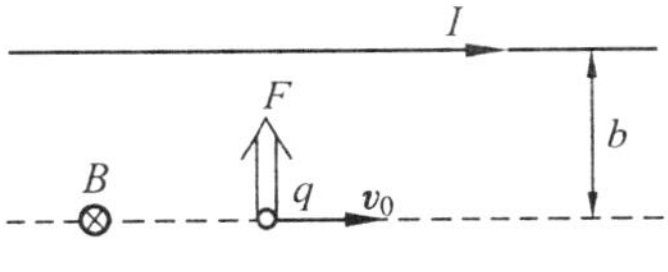


**Fig.9.11**

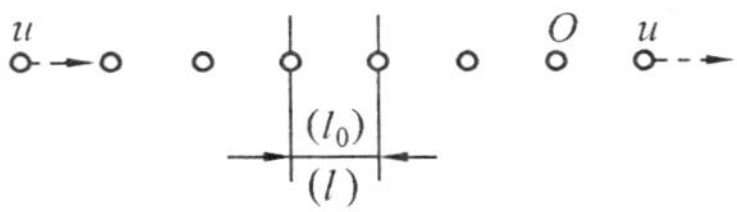


**Fig.9.12**

According to Eq.(9.36), the magnetic force acting on a charge in the case being considered is

$$F = qv_0B = qv_0 \frac{\mu_0}{4\pi}\frac{2I}{b} \tag{9.47}$$

(the meaning of the symbols is clear from Fig. 9.11). The force is directed toward the conductor carrying the current ($q > 0$).

Before commencing to derive Eq. (9.47) for the force on the basis of Coulomb's law and relativistic relations, let us consider the following effect. Assume that we have an infinite linear train of point charges of an identical magnitude $e$ spaced a very small distance $l_0$ apart (Fig.9.12). Owing to the smallness of $l_0$, we can speak of the linear density of the charges $\lambda_0$ which obviously is

$$\lambda_0 = \frac{e}{l_0} \tag{9.48}$$

Let us bring the charges into motion along the train with the identical velocity $u$. The distance between the charges will therefore diminish and become equal to

$$l = l_0\sqrt{1 - \frac{u^2}{c^2}}$$

The magnitude of the charges owing to their invariance. however, remains the same. As a result, the linear density of the charges observed in the reference frame relative to which the charges are moving will change and become equal to

$$\lambda = \frac{e}{l} = \frac{\lambda_0}{\sqrt{1 - u^2/c^2}} \tag{9.49}$$

Now let us consider in the reference frame $K$ two infinite trains formed by charges of the same magnitude, but of opposite signs, moving in opposite directions with the same velocity $u$ and virtually coinciding with each other (Fig.9.13 a). The combination of these trains is equivalent to an infinite line current having the value

$$I = 2\lambda\mu = \frac{2\lambda_0\mu}{\sqrt{1 - u^2/c^2}} \tag{9.50}$$

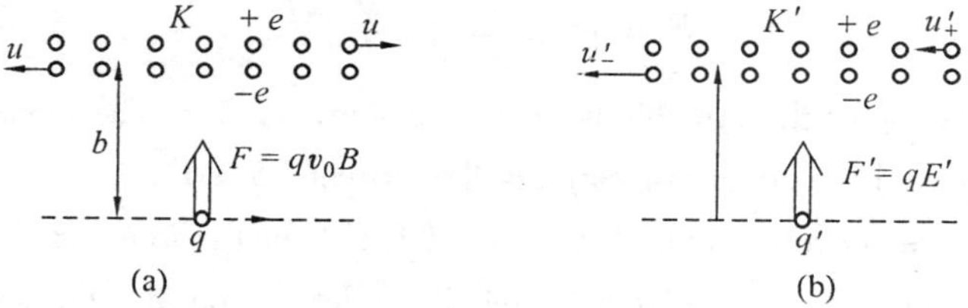


**Fig.9.13**

where $\lambda$ is the quantity determined by Eq. (9.49). The total linear density of the charges of a train equals zero, therefore an electric field is absent. The charge $q$ experiences a magnetic force whose magnitude according to Eqs. (9.47) and (9.50) is

$$F = qv_0\frac{\mu_0}{4\pi}\frac{4\lambda_0\mu}{b\sqrt{1 - u^2/c^2}} \tag{9.51}$$

Let us pass over to the reference frame $K'$ relative to which the charge $q$ is at rest (Fig.9.13b). In this frame, the charge $q$ also experiences a force (let us denote it by $F'$). This force cannot be of a magnetic origin, however, because the charge $q$ is stationary. The force $F'$ has a purely electrical origin. It appears because the linear densities of the positive and negative charges in the trains are now different (we shall see below that the density of the negative charges is greater). The surplus negative charge distributed over a train sets up an electric field that acts on the positive charge $q$ with the force $F'$ directed toward the train (see Fig.9.13b).

Let us calculate the force $F'$ and convince ourselves that it "equals" the force $F$ determined by Eq. (9.51). We have taken the word "equals" in quotation marks because force is not an invariant quantity. Upon transition from one inertial reference frame to another, the force transforms according to a quite complicated law. In a particular case, when the force $\boldsymbol{F}'$ is perpendicular to the relative velocity of the frames $K$ and $K'$ ($\boldsymbol{F}' \perp v_0$), the transformation has the form

$$\boldsymbol{F} = \frac{\boldsymbol{F}'\sqrt{1 - v_0^2/c^2} + \boldsymbol{v}_0(\boldsymbol{F}'\boldsymbol{v}')/c^2}{1 + \boldsymbol{v}_0\boldsymbol{v}'/c^2}$$

($\boldsymbol{v}'$ is the velocity of a particle experiencing the force $\boldsymbol{F}'$ and measured in the frame $K'$). If $\boldsymbol{v}' = 0$ (which occurs in the problem we are considering), the formula for transformation of the force is as follows:

$$\boldsymbol{F} = \boldsymbol{F}'\sqrt{1 - \frac{v_0^2}{c^2}}$$

A glance at this formula shows that the force perpendicular to $\boldsymbol{v}_0$ exerted on a particle at rest in the frame $K'$ is also perpendicular to the vector $\boldsymbol{v}_0$ in the frame $K$. The magnitude of the force in this case, however, is transformed by the formula

$$F = F'\sqrt{1 - \frac{v_0^2}{c^2}} \qquad (9.52)$$

The densities of the charges in the positive and negative trains measured in the frame $K'$ have the values [see Eq. (9.49)]

$$\lambda'_+ = \frac{\lambda_0}{\sqrt{1 - u'^2_+ /c^2}}, \quad \lambda'_- = \frac{\lambda_0}{\sqrt{1 - u'^2_- /c^2}} \tag{9.53}$$

where $u'_+$ and $u'_-$ are the velocities of the charges $+e$ and $-e$ measured in the frame $K'$. Upon a transition from the frame $K$ to the frame $K'$, the projection of the velocity of a particle onto the direction x coinciding with the direction of $v_0$ is transformed by the equation

$$u'_x = \frac{u_x - v_0}{1 - u_x v_0 / c^2}$$

(we have substituted $u$ and $u'$ for $v$ and $v'$). For the charges $+e$, the component $u_x$ equals $u$, for the charges $-e$ it equals $-u$ (see Fig. 9.13a). Hence

$$(u'_x)_+ = \frac{u - u_0}{1 - uv_0/c^2}, \quad (u'_x)_- = \frac{-u - v_0}{1 + uv_0/c^2}$$

Since the remaining projections equal zero, we get

$$u'_+ = \frac{|u - v_0|}{1 - uv/c^2}, \quad u'_- = \frac{u + v_0}{1 + uv/c^2} \tag{9.54}$$

To simplify our calculations, let us pass over to relative velocities:

$$\beta_0 = \frac{v_0}{c}, \quad \beta = \frac{u}{c}, \quad \beta'_+ = \frac{u'_+}{c}, \quad \beta'_- = \frac{u'_-}{c}$$

Equations (9.53) and (9.54) therefore acquire the form

$$\lambda_+ = \frac{\lambda_0}{\sqrt{1 - \beta'^2_+}}, \quad \lambda'_- = -\frac{\lambda_0}{\sqrt{1 - \beta'^2_-}} \tag{9.55}$$

$$\beta'_+ = \frac{|\beta - \beta_0|}{1 - \beta\beta_0}, \quad \beta'_- = \frac{\beta + \beta_0}{1 + \beta\beta_0} \tag{9.56}$$

With account taken of these equations, we get the following expression for the total density of the charges:

$$\lambda' = \lambda'_{+} + \lambda'_{-} = \frac{\lambda_0}{\sqrt{1 - (\frac{\beta - \beta_0}{1 - \beta\beta_0})^2}} - \frac{\lambda_0}{\sqrt{1 - (\frac{\beta + \beta_0}{1 + \beta\beta_0})^2}} =$$

$$\frac{\lambda_0(1 - \beta\beta_0)}{\sqrt{(1 - \beta\beta_0)^2 - (\beta - \beta_0)^2}} - \frac{\lambda_0(1 + \beta\beta_0)}{\sqrt{(1 + \beta\beta_0)^2 - (\beta + \beta_0)^2}}$$

It is easy to see that

$$(1 - \beta\beta_0)^2 - (\beta - \beta_0)^2 = (1 + \beta\beta_0)^2 - (\beta + \beta_0)^2 = (1 - \beta_0^2)(1 - \beta^2)$$

Consequently,

$$\lambda' = \frac{-2\lambda_0\beta\beta_0}{\sqrt{(1 - \beta_0^2)(1 - \beta^2)}} = \frac{-2\lambda_0 u v_0}{c^2\sqrt{1 - v_0^2/c^2}\sqrt{1 - u^2/c^2}} \tag{9.57}$$

An infinitely long filament carrying a charge of density $\lambda'$ sets up a field whose strength at the distance $b$ from the filament is

$$E' = \frac{1}{2\pi\varepsilon_0}\frac{\lambda'}{b}$$

In this field, the charge $q$ experiences the force

$$F' = qE' = \frac{q\lambda'}{2\pi\varepsilon_0 b}$$

Introduction of Eq. (9.57) yields (we have omitted the minus sign)

$$F' = \frac{q\lambda_0 u v_0}{\pi\varepsilon_0 b c^2\sqrt{1 - v_0^2/c^2}\sqrt{1 - u^2/c^2}} =$$

$$q v_0 \frac{\mu_0}{4\pi}\frac{4\lambda_0\mu}{b\sqrt{1 - u^2/c^2}}\frac{1}{\sqrt{1 - v_0^2/c^2}} \tag{9.58}$$

[we remind our reader that $\mu_0 = 1/\varepsilon_0 c^2$; see Eq. (9.15)]. The expression obtained differs from Eq. (9.51) only in the factor $1/\sqrt{1 - v_0^2/c^2}$. We can therefore write that

$$\boldsymbol{F} = \boldsymbol{F}'\sqrt{1 - \frac{v_0^2}{c^2}}$$

where $\boldsymbol{F}$ is the force determined by Eq. (9.51), and $\boldsymbol{F}'$ is the force determined by Eq. (9.58). A comparison with Eq. (9.52) shows that $F$ and $F'$ are the values of the same force determined in the frames $K$ and $K'$.

We must note that in the frame $K''$ which would move relative to the frame $K$ with a velocity differing from that of the charge $v_0$ the force exerted on the charge would consist of both electric and magnetic forces.

The results we have obtained signify that an electric and a magnetic field are inseparably linked with each other and form a single electromagnetic field. Upon a special choice of the reference frame, a field may be either purely electric or purely magnetic. Relative to other reference frames, however, the same field is a combination of an electric and a magnetic field.

In different inertial reference frames, the electric and magnetic fields of the same collection of charges are different. A derivation beyond the scope of a general course in physics leads to the following equations for the transformation of fields when passing over from a reference frame $K$ to a reference frame $K'$ moving relative to it with the velocity $\boldsymbol{v}_0$:

$$\left.\begin{aligned} E'_x &= E_x, E'_y = \frac{E_y - v_0 B_z}{\sqrt{1-\beta^2}}, E'_z = \frac{E_z + v_0 B_y}{\sqrt{1-\beta^2}} \\ B'_x &= B_x, B'_y = \frac{B_y + v_0 E_z}{\sqrt{1-\beta^2}}, B'_z = \frac{B_z - v_0 E_y}{\sqrt{1-\beta^2}} \end{aligned}\right\} \quad (9.59)$$

Here $E_x$, $E_y$, $E_z$, $B_x$, $B_y$, $B_z$, are the components of the vectors $\boldsymbol{E}$ and $\boldsymbol{B}$ characterizing an electromagnetic field in the frame $K$, similar primed symbols are the components of the vectors $\boldsymbol{E}'$ and $\boldsymbol{B}'$ characterizing the field in the frame $K'$. The Greek letter $\beta$ stands for the ratio $v_0/c$.

Resolving the vectors $\boldsymbol{E}$ and $\boldsymbol{B}$, and also $\boldsymbol{E}'$ and $\boldsymbol{B}'$, into their components parallel to the vector $v_0$ (and, consequently, to the axes $x$ and $x'$) and perpendicular to this vector (i.e. representing, for example, $\boldsymbol{E}$ in the form $\boldsymbol{E} = \boldsymbol{E}_{\parallel} + \boldsymbol{E}_{\perp}$, etc.), we can write Eq.(9.59) in the vector form:

$$\left.\begin{aligned} \boldsymbol{E}'_{\parallel} &= \boldsymbol{E}_{\parallel}, \boldsymbol{E}'_{\perp} = \frac{\boldsymbol{E}_{\perp} + [v_0 \boldsymbol{B}_{\perp}]}{\sqrt{1-\beta^2}} \\ \boldsymbol{B}'_{\parallel} &= \boldsymbol{B}_{\parallel}, \boldsymbol{B}'_{\perp} = \frac{\boldsymbol{B}_{\perp} - (1/c^2)[v_0 \boldsymbol{E}_{\perp}]}{\sqrt{1-\beta^2}} \end{aligned}\right\} \tag{9.60}$$

In the Gaussian system of units, Eq. (9.60) have the form

$$\left.\begin{aligned} \boldsymbol{E}'_{\parallel} &= \boldsymbol{E}_{\parallel}, \boldsymbol{E}'_{\perp} = \frac{\boldsymbol{E}_{\perp} + (1/c)[v_0 \boldsymbol{B}_{\perp}]}{\sqrt{1-\beta^2}} \\ \boldsymbol{B}'_{\parallel} &= \boldsymbol{B}_{\parallel}, \boldsymbol{B}'_{\perp} = \frac{\boldsymbol{B}_{\perp} - (1/c)[v_0 \boldsymbol{E}_{\perp}]}{\sqrt{1-\beta^2}} \end{aligned}\right\} \tag{9.61}$$

When $\beta \ll 1$ (i.e. $v_0 \ll c$), Eq. (9.60) are simplified as follows:

$$\boldsymbol{E}'_{\parallel} = \boldsymbol{E}_{\parallel}, \boldsymbol{E}'_{\perp} = \boldsymbol{E}_{\perp} + [v_0 \boldsymbol{B}_{\perp}]$$

$$\boldsymbol{B}'_{\parallel} = \boldsymbol{B}_{\parallel}, \boldsymbol{B}'_{\perp} = \boldsymbol{B}_{\perp} - \frac{1}{c^2}[v_0 \boldsymbol{E}_{\perp}]$$

Adding these equations in pairs, we get

$$\begin{aligned} \boldsymbol{E}' &= \boldsymbol{E}'_{\parallel} + \boldsymbol{E}'_{\perp} = \boldsymbol{E}_{\parallel} + \boldsymbol{E}_{\perp} + [v_0 \boldsymbol{B}_{\perp}] = \boldsymbol{E} + [v_0 \boldsymbol{B}_{\perp}] \\ \boldsymbol{B}' &= \boldsymbol{B}'_{\parallel} + \boldsymbol{B}'_{\perp} = \boldsymbol{B}_{\parallel} + \boldsymbol{B}_{\perp} - \frac{1}{c^2}[v_0 \boldsymbol{E}_{\perp}] = \boldsymbol{B} - \frac{1}{c^2}[v_0 \boldsymbol{E}_{\perp}] \end{aligned} \tag{9.62}$$

Since the vectors $\boldsymbol{v}_0$ *and* $\boldsymbol{B}_{\parallel}$ *are collinear, their vector product equals zero. Hence*, $[\boldsymbol{v}_0 \boldsymbol{B}] = [\boldsymbol{v}_0 \boldsymbol{B}_{\parallel}] + [\boldsymbol{v}_0 \boldsymbol{B}_{\perp}] = [\boldsymbol{v}_0 \boldsymbol{B}_{\perp}]$. Similarly, $[\boldsymbol{v}_0 \boldsymbol{E}] = [\boldsymbol{v}_0 \boldsymbol{E}_{\perp}]$. With this taken into account, Eq.(9.62) can be given the form

$$\boldsymbol{E}' = \boldsymbol{E} + [V_0 B], \quad \boldsymbol{B}' = \boldsymbol{B} - \frac{1}{c^2}[v_0 \boldsymbol{E}] \tag{9.63}$$

Fields are transformed by means of these equations if the relative velocity of the reference frames $v_0$ is much smaller than the speed of light in a vacuum $c$ ($v_0 \ll c$).

Equations (9.63) acquire the following form in the Gaussian system of units:

$$E' = E + \frac{1}{c}[v_0 B], B' = B - \frac{1}{c}[v_0 E] \tag{9.64}$$

In the example in the frame $K$ considered at the beginning of this Section in which the charge $q$ travelled with the velocity $\boldsymbol{v}_0$ parallel to a current - carrying wire, there was only the magnetic field $\boldsymbol{B}_\perp$ perpendicular to $\boldsymbol{v}_0$; the components $\boldsymbol{B}_\parallel$, $\boldsymbol{E}_\perp$ and $\boldsymbol{E}_\parallel$ equaled zero. According to Eq. (9.60) in the frame $K'$ in which the charge $q$ is at rest (this frame travels relative to $K$ with the velocity $\boldsymbol{v}_0$), the component $\boldsymbol{B}'_\perp$ equal to $\boldsymbol{B}_\perp\sqrt{1-\beta^2}$ is observed and, in addition, the perpendicular component of the electric field $\boldsymbol{E}'_\perp = [v_0 \boldsymbol{B}_\perp]\sqrt{1-\beta^2}$.

In the frame $K$, the charge experiences the force

$$\boldsymbol{F} = q[\boldsymbol{v}_0 \boldsymbol{B}_\perp] \tag{9.65}$$

Since the charge $q$ is rest in the frame $K'$, it experiences in this Frame only the electric force

$$\boldsymbol{F}' = q\boldsymbol{E}'_\perp = \frac{q[\boldsymbol{v}_0 \boldsymbol{B}_\perp]}{\sqrt{1-\beta^2}} \tag{9.66}$$

A comparison of Eqs. (9.65) and (9.66) yields $\boldsymbol{F} = \boldsymbol{F}'\sqrt{1-\beta^2}$ Which coincides with Eq. (9.52).

## Summary of Key Terms

**Magnetic force** (1) Between magnets, it is the attraction of unlike magnetic poles for each other and the repulsion between like magnetic poles. (2) Between magnets field and a moving charge, it is a deflecting force due to the motion of the charge: the deflecting force is perpendicular to the motion of the charge and perpendicular to the magnetic field lines. This force is greatest when the charge moves perpendicular to the field lines and is smallest when moving parallel to the field lines.

**Magnetic field** The region of magnetic influence around a magnetic pole or a moving charged particle. Magnetic domains Clustered regions of aligned magnetic atoms. When these regions themselves are aligned with one another, the substance containing them is a magnet.

**Electromagnet** A magnet whose field is produced by an electric current. Usually in the form of a wire coil with a piece of iron inside the coil.

10

# Maxwell's Equations

## 10.1 VORTEX ELECTRIC FIELD

Let us consider electromagnetic induction when a wire loop in which a current is induced is stationary, and the changes in the magnetic flux are due to changes in the magnetic field. The setting up of an induced current signifies that the changes in the magnetic field produce extraneous forces in the loop that are exerted on the current carriers. These extraneous forces are associated neither with chemical nor with thermal processes in the wire. They also cannot be magnetic forces because such forces do not work on charges. It remains to conclude that the induced current is due to the electric field set up in the wire. Let us use the symbol $\boldsymbol{E}_B$ to denote the strength of this field (this symbol, like the one $\boldsymbol{E}_q$ used below, is an auxiliary one; we shall omit the subscripts $B$ and $q$ later on). The e. m. f. equals the circulation of the vector $\boldsymbol{E}_B$ around the given loop:

$$\varepsilon_i = \oint \boldsymbol{E}_B \mathrm{d}\boldsymbol{l} \tag{10.1}$$

Introducing into the equation $\varepsilon_i = -\mathrm{d}\Phi/\mathrm{d}t$ Eq. (10.1) for $\varepsilon_i$ and the expression $\int \boldsymbol{B}\mathrm{d}\boldsymbol{S}$ for $\Phi$, we arrive at the equation

$$\oint \boldsymbol{E}_B \mathrm{d}\boldsymbol{l} = -\frac{\mathrm{d}}{\mathrm{d}t}\int_S \boldsymbol{B}\mathrm{d}\boldsymbol{S}$$

(the integral in the right-hand side of the equation is taken over an arbitrary surface resting on the loop). Since the loop and the surface are stationary, the operations of time differentiation and integration over the surface can have their places exchanged:

$$\oint \boldsymbol{E}_B \mathrm{d}\boldsymbol{l} = -\int_B \frac{\partial \boldsymbol{B}}{\partial t}\mathrm{d}\boldsymbol{S} \qquad (10.2)$$

In connection with the fact that the vector $\boldsymbol{B}$ depends, generally speaking, both on the time and on the coordinates, we have put the symbol of the partial time derivative inside the integral (the integral $\int \boldsymbol{B}\mathrm{d}\boldsymbol{S}$ is a function only of time).

Let us transform the left-hand side of Eq. (10.2) in accordance with Stokes's theorem. The result is

$$\int_S [\nabla \boldsymbol{E}_B]\mathrm{d}\boldsymbol{S} = -\int_S \frac{\mathrm{d}\boldsymbol{B}}{\mathrm{d}t}\mathrm{d}\boldsymbol{S}$$

Owing to the arbitrary nature of choosing the integration surface, the following equation must be obeyed:

$$[\nabla \boldsymbol{E}_B] = -\frac{\partial \boldsymbol{B}}{\partial t} \qquad (10.3)$$

The curl of the field $\boldsymbol{E}_B$ at each point of space equals the time derivative of the vector $\boldsymbol{B}$ taken with the opposite sign.

The British physicist James Maxwell (1831 ~ 1879) assumed that a time – varying magnetic field causes the field $\boldsymbol{E}_B$ to appear in space regardless of whether or not there is a wire loop in this space. The presence of a loop only makes it possible to detect the existence of an electric field at the corresponding points of space as a result of a current being induced in the loop.

Thus according to Maxwell's idea, a time-varying magnetic field gives birth to an electric field. This field $\boldsymbol{E}_B$ differs appreciably from the electrostatic field $\boldsymbol{E}_q$ set up by fixed charges. An electrostatic field is a potential one, its strength lines begin and terminate at charges. The curl of the vector $\boldsymbol{E}_q$ is zero at any point:

$$[\nabla \boldsymbol{E}_q] = 0 \tag{10.4}$$

According to Eq. (10.3), the curl of the vector $\boldsymbol{E}_B$ differs from zero. Hence, the field $\boldsymbol{E}_B$ like a magnetic field, is a vortex one. The strength lines of the field $\boldsymbol{E}_B$ are closed.

Thus, an electric field may be either a potential ($\boldsymbol{E}_q$) or a vortex ($\boldsymbol{E}_B$) one. In the general case, an electric field can consist of the field $\boldsymbol{E}_q$ produced by charges and the field $\boldsymbol{E}_q$ set up by a time varying field. Adding Eqs. (10.4) and (10.3), we get the following equation for the curl of the strength of the total field $\boldsymbol{E} = \boldsymbol{E}_q + \boldsymbol{E}_B$:

$$[\nabla \boldsymbol{E}] = -\frac{\partial \boldsymbol{B}}{\partial t} \tag{10.5}$$

This equation is one of the fundamental ones in Maxwell's electroMagnetic theory.

The existence of a relationship between electric and magnetic fields [expressed in particular by Eq. (10.5)] is a reason why the separate treatment of these fields has only a relative meaning. Indeed, an electric field is set up by a system of fixed charges. If the charges are fixed relative to a certain inertial reference frame, however, they are moving relative to other inertial frames and, consequently, set up not only an electric, but also a magnetic field. A stationary wire carrying a steady current sets up a constant magnetic field at every point of space. This wire is in motion, however, relative to other inertial frames. Consequently, the magnetic field it sets up at any point with the given coordinates $x$, $y$, $z$ will change and therefore give birth to a vortex electric field. Thus, a field which is "purely" electric or "purely" magnetic relative to a certain reference frame will be a combination of an electric and a magnetic field forming a single electromagnetic field relative to other reference frames.

## 10.2 DISPLACEMENT CURRENT

For a stationary (i.e. not varying with time) electromagnetic field, the curl of the vector $\boldsymbol{H}$ by $\nabla \boldsymbol{H} = \boldsymbol{j}$ equals the density of the conduction current at each point:

$$[\nabla \boldsymbol{H}] = \boldsymbol{j}$$

The vector $\boldsymbol{j}$ is associated with the charge density at the same point. By continuity equation (10.11):

$$\nabla \boldsymbol{j} = -\frac{\partial \rho}{\partial t}$$

An electromagnetic field can be stationary only provided that the charge density $\rho$ and the current density $j$ do not depend on the time. In this case, the divergence of $j$ equals zero. Therefore, the current lines (lines of the vector $j$) have no sources and are closed.

Let us see whether $\nabla \boldsymbol{H} = \boldsymbol{j}$ holds for time-varying fields. We shall consider the current flowing when a capacitor is charged from a source of constant voltage $U$. This current varies with time (the current stops flowing when the voltage across the capacitor becomes equal to $U$). The lines of the conduction current are interrupted in the space between the capacitor plates (Fig. 10.1; the current lines inside the plates are shown by dash lines).

Let us take a circular loop $\Gamma$ enclosing the wire in which the current flows toward the capacitor and integrate $\nabla \boldsymbol{H} = \boldsymbol{j}$ over surface $S_1$ intersecting the wire and enclosed by the loop:

$$\int_{S_1} [\nabla \boldsymbol{H}] \mathrm{d}S = \int_{S_1} \boldsymbol{j} \mathrm{d}S$$

Transforming the left-hand side according to Stokes's theorem we get the circulation of the vector $\boldsymbol{H}$ over loop $\Gamma$:

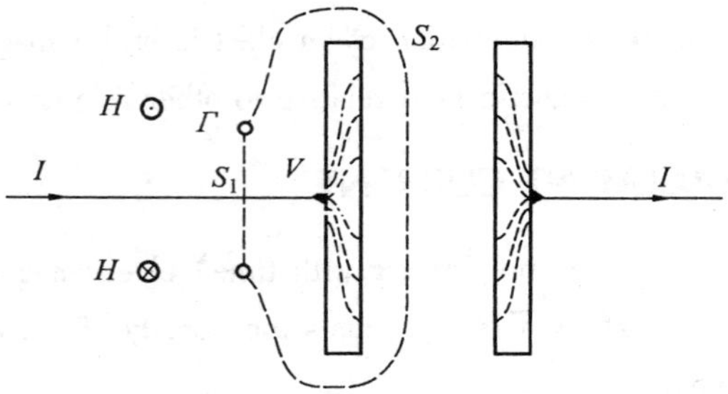


**Fig. 10.1**

$$\int_{\Gamma} \boldsymbol{H} \mathrm{d}\boldsymbol{l} = \int_{S_1} \boldsymbol{j} \mathrm{d}S = I \qquad (10.6)$$

($I$ is the current charging the capacitor). After performing similar calculations for surface $S_2$ that does not intersect the current-carrying wire (see Fig. 10.1), we arrive at the obviously incorrect relation

$$\int_{\Gamma} \boldsymbol{H} \mathrm{d}\boldsymbol{l} = \int_{S_2} \boldsymbol{j} \mathrm{d}\boldsymbol{S} = 0 \qquad (10.7)$$

The result we have obtained indicates that for time-varying fields $\nabla \boldsymbol{H} = \boldsymbol{j}$ stops being correct. The conclusion suggests itself that this equation lacks an addend depending on the time derivatives of the fields. For stationary fields, this addend vanishes.

That $\nabla \boldsymbol{H} = \boldsymbol{j}$ is not correct for non-stationary fields is also indicated by the following reasoning. Let us take the divergence of both sides of $\nabla \boldsymbol{H} = \boldsymbol{j}$:

$$\nabla [\nabla \boldsymbol{H}] = \nabla \boldsymbol{j}$$

The divergence of a curl must equal zero. We thus arrive at the conclusion that the divergence of the vector $\boldsymbol{j}$ must also always equal zero. But this conclusion contradicts the continuity $\nabla \boldsymbol{j} = -\frac{\partial \rho}{\partial t}$. Indeed, in non-stationary processes, $\rho$ may change with time (this, in particular, is what happens with the charge density on the plates of a capacitor being charged). In this case in accordance with $\nabla \boldsymbol{j} = -\frac{\partial \rho}{\partial t}$, the divergence of $\boldsymbol{j}$ diners from zero.

To bring $\nabla \boldsymbol{H} = \boldsymbol{j}$ and $\nabla \boldsymbol{j} = -\frac{\partial \rho}{\partial t}$ into agreement Maxwell introduced an additional addend into the right-hand side of $\nabla \boldsymbol{H} = \boldsymbol{j}$. It is quite natural that this addend should have the dimension of current density. Maxwell called it the density of the displacement current. Thus, according to Maxwell, $\nabla \boldsymbol{H} = \boldsymbol{j}$ should have the form

$$[\nabla \boldsymbol{H}] = \boldsymbol{j} + \boldsymbol{j}_{\mathrm{d}} \tag{10.8}$$

The sum of the conduction current and the displacement current is usually called the total current. The density of the total current is

$$\boldsymbol{j}_{\mathrm{tot}} = \boldsymbol{j} + \boldsymbol{j}_{\mathrm{d}} \tag{10.9}$$

If we assume that the divergence of the displacement current equals that of the conduction current taken with the opposite sign:

$$\nabla \boldsymbol{j}_{\mathrm{d}} = -\nabla \boldsymbol{j} \tag{10.10}$$

then the divergence of the right-hand side of Eq. (10.8), like that of the left-hand side, will always be zero.

Substituting $\partial \rho / \partial t$ for $\nabla \boldsymbol{j}$ in Eq. (10.10) in accordance with $\nabla \boldsymbol{j} = -\frac{\partial \rho}{\partial t}$ we get the following expression for the divergence of the displacement current:

$$\nabla \boldsymbol{j}_{\mathrm{d}} = \frac{\partial \rho}{\partial t} \tag{10.11}$$

To associate the displacement current with quantities characterizing the change in an electric field with time, let us use $\nabla \boldsymbol{D} = \rho$ according to which the divergence of the electric displacement vector equals the density of the extraneous charges:

$$\nabla \boldsymbol{D} = \rho$$

Time differentiation of this equation yields

$$\frac{\partial}{\partial t}(\nabla \boldsymbol{D}) = \frac{\partial \rho}{\partial t}$$

Now let us change the sequence of differentiation with respect to time and to the coordinates in the left-hand side. As a result, we get the following expression for the derivative of $\rho$ with respect to $t$:

$$\frac{\partial \rho}{\partial t} = \nabla \left( \frac{\partial \boldsymbol{D}}{\partial t} \right)$$

Introduction of this expression into Eq.(10.11) yields

$$\nabla \boldsymbol{j}_{\mathrm{d}} = \nabla \left( \frac{\partial \boldsymbol{D}}{\partial t} \right)$$

Hence

$$\boldsymbol{j}_{\mathrm{d}} = \frac{\partial \boldsymbol{D}}{\partial t} \qquad (10.12)$$

Using Eq.(10.12) in Eq.(10.8), we arrive at the equation

$$[\nabla \boldsymbol{H}] = \boldsymbol{j} + \frac{\partial \boldsymbol{D}}{\partial t} \qquad (10.13)$$

which, like Eq. (10.5), is one of the fundamental equations in Maxwell's theory.

We must underline the fact that the term "displacement current" is purely conventional. In essence, the displacement current is a time-varying electric field. The only reason for calling the quantity given by Eq. (10.12) a "current" is that the dimension of this quantity coincides with that of current density. Of all the physical properties belonging to a real current, a displacement current has only one—the ability of producing a magnetic field.

The introduction of the displacement current determined by Eq. (10.12) has "given equal rights" to an electric field and a magnetic field. It can be seen from the phenomenon of electromagnetic induction that a varying magnetic field sets up an electric field. It follows from Eq. (10.13) that a varying electric field sets up a magnetic field.

There is a displacement current wherever there is a time-varying electric field. In particular, it also exists inside conductors carrying an alternating electric current. The displacement current inside conductors, however, is usually negligibly small in comparison with the conduction current.

We must note that Eq. (10.6) is approximate. For it to become quite strict, we must add a term to its right-hand side that takes account of the displacement current due to the weak dispersed electric field in the vicinity of surface $S_1$.

Let us convince ourselves that the surface integral of the right-hand side of Eq. (10.8) has the same value for surfaces $S_1$ and $S_2$, (see Fig. 10.1). Both the conduction current and the displacement current due to the electric field outside the capacitor "flow" through surface $S_1$. Hence, for the first surface, we have

$$Int_1 = \int_{S_1} \boldsymbol{j} \mathrm{d}\boldsymbol{S} + \frac{\mathrm{d}}{\mathrm{d}t}\int_{S_1} \boldsymbol{D} \mathrm{d}\boldsymbol{S} = I + \frac{\mathrm{d}}{\mathrm{d}t}\Phi_{1,\mathrm{in}}$$

(we have changed the sequence of the operations of differentiation with respect to time and integration over the coordinates in the second addend). The quantity designated by the letter $I$ is the current flowing in the conductor to the left-hand plate of the capacitor, $\Phi_{1,\mathrm{in}}$ is the flux of the vector $\boldsymbol{D}$ flowing into the volume $V$ bounded by surfaces $S_1$ and $S_2$, (see Fig. 10.1).

For the second surface, $\boldsymbol{j} = 0$, consequently

$$Int_2 = \frac{\mathrm{d}}{\mathrm{d}t}\int_{S_2} \boldsymbol{D} \mathrm{d}\boldsymbol{S} = \frac{\mathrm{d}}{\mathrm{d}t}\Phi_{2,\mathrm{out}}$$

where $\Phi_{2,\mathrm{out}}$ out is the flux of the vector $\boldsymbol{D}$ flowing out of volume $V$ through surface $S_2$.

The difference between the integrals is

$$Int_2 - Int_1 = \frac{\mathrm{d}}{\mathrm{d}t}\Phi_{2,\mathrm{out}} - \frac{\mathrm{d}}{\mathrm{d}t}\Phi_{1,\mathrm{in}} - I$$

The current $I$ can be represented as $\mathrm{d}q/\mathrm{d}t$, where $q$ is the charge on a capacitor plate. The flux parsing inward through surface $S_1$ equals the flux passing outward through the same surface taken with the opposite sign. Substituting $-\Phi_{1,\mathrm{out}}$ for $\Phi_{1,\mathrm{in}}$ and $\mathrm{d}q/\mathrm{d}t$ for $I$, we get

$$Int_2 - Int_1 = \frac{\mathrm{d}}{\mathrm{d}t}(\Phi_{2,\mathrm{out}} + \Phi_{1,\mathrm{out}}) - \frac{\mathrm{d}q}{\mathrm{d}t} = \frac{\mathrm{d}}{\mathrm{d}t}(\Phi_D - q) \tag{10.14}$$

where $\Phi_D$ is the flux of the vector $\boldsymbol{D}$ through the closed surface formed by surfaces $S_1$, and $S_2$,. This flux must equal the charge enclosed by the surface. In the given case, it is the charge $q$ on a capacitor plate. Thus, the right-hand side of Eq. (10.14) equals zero. It follows that the magnitude of the surface integral of the total current density vector does not depend on the choice of the surface over which the integral is being calculated.

We can construct current lines for the displacement current like those for the conduction current. The electric displacement in the space between the capacitor plates equals the surface charge density on a plate: $D = \sigma$. Hence, $\dot{D} = \dot{\sigma}$. The left-hand side gives the density of the displacement current in the space between the plates, and the right-hand side—the density of the conduction current inside the plates. The equality of these densities signifies that the lines of the conduction current uninterruptedly transform into lines of the displacement current at the boundary of the plates. Consequently, the lines of the total current are closed.

## 10.3 MAXWELL'S EQUATIONS

The discovery of the displacement current permitted Maxwell to present a single general theory of electrical and magnetic phenomena. This theory explained all the experimental facts known at that time and predicted a number of new phenomena whose existence was confirmed later on. The main corollary of Maxwell's theory was the conclusion on the existence of electromagnetic waves propagating with the speed of light. Theoretical investigation of the properties of these waves led Maxwell to the electromagnetic theory of light.

The theory is based on Maxwell's equations. These equations play the same part in the science of electromagnetism as Newton's laws do in mechanics, or the fundamental laws in thermodynamics.

The first pair of Maxwell's equations is formed by Eqs. (10.5) and (d):

$$[\nabla \boldsymbol{E}] = -\frac{\partial \boldsymbol{B}}{\partial t} \tag{10.5}$$

$$\nabla \boldsymbol{B} = 0 \tag{d}$$

The first of them relates the values of $\boldsymbol{E}$ to changes in the vector $\boldsymbol{B}$ in time and is in essence an expression of the law of electromagnetic Induction. The second one points to the absence of sources of a magnetic field, i.e. magnetic charges.

The second pair of Maxwell's rations is formed by Eqs. (10.13) and (c)

$$[\nabla \boldsymbol{H}] = \boldsymbol{j} + \frac{\partial \boldsymbol{D}}{\partial t} \tag{10.13}$$

$$\nabla \boldsymbol{D} = \rho \tag{c}$$

The first of them establishes a relation between the conduction and displacement currents and the magnetic field they produce. The second one shows that extraneous charges are the sources of the vector $\boldsymbol{D}$.

Equations (10.5), (d), (10.13) and (c) are Maxwell's equations In the differential form. We must note that the first pair of equations. includes only the basic characteristics of a field, namely, $\boldsymbol{E}$ and $\boldsymbol{B}$. The second pair includes only the auxiliary quantities $\boldsymbol{D}$ and $\boldsymbol{H}$.

Each of the vector equations (10.5) and (10.13) is equivalent to three scalar equations relating the components of the vectors in the left-hand and right-hand sides of the equations. Let us present Maxwell's equation in the scalar form:

$$\left.\begin{aligned}\frac{\partial E_z}{\partial y}-\frac{\partial E_Y}{\partial z}&=-\frac{\partial B_x}{\partial t}\\ \frac{\partial E_x}{\partial z}-\frac{\partial E_z}{\partial x}&=-\frac{\partial B_Y}{\partial t}\\ \frac{\partial E_Y}{\partial x}-\frac{\partial E_x}{\partial y}&=-\frac{\partial B_z}{\partial t}\end{aligned}\right\}\qquad(10.15)$$

$$\frac{\partial B_x}{\partial x}+\frac{\partial B_Y}{\partial y}+\frac{\partial B_z}{\partial z}=0\qquad(10.16)$$

(the first pair of equations),

$$\left.\begin{aligned}\frac{\partial H_z}{\partial y}-\frac{\partial H_Y}{\partial z}&=j_x+\frac{\partial D_x}{\partial t}\\ \frac{\partial H_x}{\partial z}-\frac{\partial H_z}{\partial x}&=j_Y+\frac{\partial D_Y}{\partial t}\\ \frac{\partial H_Y}{\partial x}-\frac{\partial H_x}{\partial y}&=j_z+\frac{\partial D_z}{\partial t}\end{aligned}\right\}\qquad(10.17)$$

$$\frac{\partial D_x}{\partial y}+\frac{\partial D_Y}{\partial y}+\frac{\partial D_z}{\partial z}=\rho\qquad(10.18)$$

the second pair of equations).

We get a total of eight equations including twelve functions (three components each of the vectors $\boldsymbol{E}$, $\boldsymbol{B}$, $\boldsymbol{D}$, $\boldsymbol{H}$). Since the number of equations is less than the number of unknown functions, Eq. (10.5), $\nabla \boldsymbol{B} = \nabla \boldsymbol{B}_0 + \nabla \boldsymbol{B}'_0 = 0$, Eq. (10.13), and $\nabla \boldsymbol{D} = \boldsymbol{P}$ are not sufficient for finding the fields according to the given distribution of the charges and currents. To calculate the fields, we must add equations relating $\boldsymbol{D}$ and $\boldsymbol{j}$ to $\boldsymbol{E}$, and also $\boldsymbol{H}$ to $\boldsymbol{B}$ to these equations. They have the form

$$\boldsymbol{D} = \varepsilon_0 \varepsilon \boldsymbol{E} \qquad (\mathrm{e})$$

$$\boldsymbol{B} = \mu_0 \mu \boldsymbol{H} \qquad (\mathrm{f})$$

$$\boldsymbol{j} = \sigma \boldsymbol{E} \qquad (\mathrm{g})$$

The collection of equations (10.5), (d), (10.13), (c), and (e), (f), (g) forms the foundation of the electrodynamics of media at rest.

The equations

$$\oint_{\Gamma} \boldsymbol{E} \mathrm{d} \boldsymbol{l} = -\frac{\mathrm{d}}{\mathrm{d}t} \int_{B} \boldsymbol{B} \mathrm{d} \boldsymbol{S} \tag{10.19}$$

$$\int_{S} \boldsymbol{B} \mathrm{d} \boldsymbol{S} = 0 \tag{10.20}$$

(the first pair) and

$$\oint_{\Gamma} \boldsymbol{H} \mathrm{d} \boldsymbol{l} = \int \boldsymbol{j} \mathrm{d} \boldsymbol{S} + \frac{\mathrm{d}}{\mathrm{d}t} \int_{S} \boldsymbol{D} \mathrm{d} \boldsymbol{S} \tag{10.21}$$

$$\oint_{S} \boldsymbol{D} \mathrm{d} \boldsymbol{S} = \int_{V} \rho \mathrm{d} V \tag{10.22}$$

(the second pair) are Maxwell's equations in the integral form.

Equation (10.19) is obtained by integration of Eq. (10.5) over arbitrary surface $S$ with the following transformation of the left-hand side according to Stokes's theorem into an integral over loop $\Gamma$ enclosing surface $S$. Equation (10.21) is obtained in the same way from Eq. (10.13). Equations (10.20) and (10.22) are obtained from Eqs. (d) and (c) by integration over the arbitrary volume $V$ with the following transformation of the left-hand side according to the Ostrogradsky-Gauss theorem into an integral over closed surface $S$ enclosing volume $V$.

## Reading Materials:

### Superconductivity

The electrical resistance of a metal conductor can be decreased by cooling. You may wonder how far one can go with this decrease. Strangely enough, for certain materials you can go all the way. At relatively low temperatures some materials exhibit superconductivity, and the electrical resistance vanishes or goes to zero!

In this state, an electric current established in a superconducting ring have been observed to remain constant for several years. In a sense, the conducting electrons of a superconductor never collide with the material lattice, while a conductor at normal temperatures always has some resistance and energy loss. The trick or requirement for superconduction is maintaining materials at very low temperatures.

Superconductivity was discovered by a Dutch physicist named Heike Onnes in 1911 and was first observed in solid mercury at a temperature of about 4 K(4 kelvins above absolute zero, −269℃ or −452°F —pretty cold!). The mercury was cooled to this temperature at which helium temperatures.

Liquid helium is relatively expensive, $3 to $6 a liter, depending on the amount you buy. So there has been a search for other materials that become superconducting at higher temperatures. Over the years, other superconducting metals and alloys were found, and the critical temperature crept upward to about 18 K(−225℃ or −427°F). In 1973, this went up to 23 K(−250℃ or −418°F).

In 1986 there was a major break through and a new class of superconductors was discovered with higher critical temperatures. These were ceramic "alloys" or mixtures of rare earth element such as lanthanum and yttrium. The new superconductors were prepared by grinding a blend of metallic elements and heating them at a high temperature, which produces a ceramic material. For example, one of the new superconductors was a mixture of yttrium, barium, and copper oxide. In 1986 the critical temperature got up to 57 K(−216℃ or −283°F), and in 1987 a critical temperature of 98 K(−175℃ or −183°F) was obtained. In the first part of 1988, during the time this book was being readied for press, a new thallium compound was reported that achieves superconductivity at 125 K(−148℃ or −234°F).

There have also been reports of bits of materials becoming superconducting at room temperature (2 950 K), but these are questioned.

The 98 K material was a major breakthrough.. Such "high temperature" superconductors can be cooled using liquid nitrogen. (see figure above.) Liquid nitrogen has a boiling point of 77 K ( − 196℃ or − 321°F ) and costs only about 25 to 40 cents per liter. ( Nitrogen is readily plentiful as the major constituent of the air.) This is an important scientific discovery that will probably revolutionize many things. However, it's going to take a while, which many reports fail to point out.

One of the main applications of superconductors is in superconducting magnets. Magnets are used in motors. And the greater the strength of the magnet, the more powerful the motor. In such electromagnets, the strength of the magnet depends on the current in the windings (wires). Without resistance, no loss occurs and there is a greater current. (Superconducring magnets have been used on ship motors for some years but with liquid helium.)

With superconducting magnets, things are more efficient and larger currents can be conducted. Potential applications are magnetically levitated trains, which are repelled off the track and ride on cushions of air. Superconducting magnets could also be used for propulsion (Experimental "MageLev" trains have already been built using low-temperature superconducting magnets.) Other applications might be underground electrical transmission cables or electric cars.

However, it is generally thought that such application are some distance in the future. A more immediate application Will probably be in computer interconnects. Interconnects are the metallic connections between computer chips by which they "talk" to each other. Superconducting interconnects would decrease power dissipation and possibly speed up the signal transfer, which makes for faster computers. The absence of electrical resistance opens up many possibilities.

# PART 5
# MODERN PHYSICS

# 11

# Relativity

## 11.1 THE BACKGROUND

The background of the theory of relativity is full of discovery and debate. For most people, the term relativity or the theory of relativity immediately brings to mind Albeit Einstein (and vice versa). Indeed, Einstein did formulate a theory of relativity, which consists of two main parts: the special theory (circa 1905) and the general theory (circa 1915). But before looking at this theory, let's set the stage for its development.

The theory of relativity involves light, in particular, the speed of light. From common experience we see that light appears to travel instantaneously from one place to another. Measuring its speed is no easy task. Galileo tried to do this by measuring the time it took light signals to travel several kilometers. Galileo's signals came from a lantern. When a covering bucket was removed, a light beam was sent to an assistant, who removed a bucket from another lantern and sent a signal back to Galileo.

As you might imagine, the experimental results were not very good (in fact, they were useless), since the assistant's and Galileo's reaction times were involved. (For typical people, this reaction time would allow light to go all the way around the Earth.) The speed of light was subsequently measured using astronomical methods.

The first successful terrestrial method of measuring the speed of light was carried out in 1849 by the French physicist H. L. Fizeau. The idea of Fizeau's cogwheel method is illustrated in Figure . If the cogwheel is not moving, then the light from a candle will go through the opening between cogs 1 and 2, travel to the mirror, be reflected back, and pass again through the opening between the cogs to an observer behind the candle. However, if the wheel is rotated, the beam is "chopped" up.

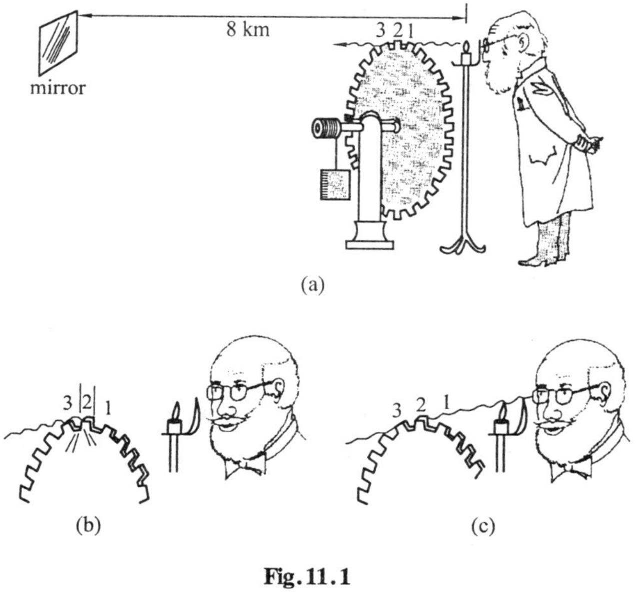


**Fig.11.1**

The speed of light. Fizeau's cogwheel method for measuring the speed of light. See text for description.

When the wheel is rotating slowly, a segment of the light beam – for example, that chopped off between cogs 1 and 2 – will be reflected back and will arrive while cog 2 is in front of the observer. But at a faster rotational speed, the cog will be out of the way and the light – beam segment will pass between cogs 2 and 3 and will be seen by the observer. The time it takes for the next cog space to rotate in front of the observer can be calculated from the rotational speed, and if we know the distance travelled by the light, the speed of light can also be calculated.

Fizeau's results were not extremely accurate, but they were pretty good considering the equipment. His experimental result for the speed of light was about 5 percent off the present – day accepted value: (speed of light) $c = 3 \times 10^8$ m/s(186 000 mi/s).

At this rate, it takes about 8 min for light to reach us from the Sun. The closest star to our solar system, Alpha Centauri, is about 4.3 light years away. A light year is the distance light can travel in a year (about 10 trillion kilometers or 5.8 trillion miles). Hence, the light we see from Alpha Centauri is more than four years "old", since it takes 4.3 years for the light to reach us. How about the light from a star 1 000 light years away?

## 11.2 THE ETHER

With the wave nature of light being demonstrated through interference experiments and the speed of light being reasonably well known by astronomical methods prior to the 1800's, scientists turned their attention to the consideration of the medium that carried or propagated light waves. From general experience, it was thought that a medium was necessary for wave propagation. For example, sound waves propagate in air (and other media). It could be shown that sound waves could not travel through a vacuum – some material substance was needed for their propagation. Similarly, you couldn't have water waves without water. By such reasoning, it was also believed that light waves had to have a medium of propagation.

How then did light propagate through the void of space, for example, from the Sun to Earth? Not being able to see how light could travel through nothing, scientists created a hypothetical medium that was called luminiferous ether, or just plain ether. This substance was presumed to occupy interstellar space and to be present in all materials through which light travels. The idea of the existence of the ether seemed so logical that it quickly gained widespread acceptance.

It is easy to postulate the existence of something, but by the scientific method, a theory must be substantiated by experiment. So scientists set out to detect the ether experimentally. This set the stage for Einstein's theory of relativity.

## 11.3 THE MICHELSON-MORLEY EXPERIMENT

Since the ether permeated all space, it was reasoned that it was the one thing that remained fixed in the universe. This belief led to an idea about how to detect the ether and prove its existence. The Earth's average speed through the stationary ether in its revolution about the Sun is about 30 km/s. Moving through the ether would cause an ether "wind" to blow over the Earth. This is similar to the "wind" you experience when travelling in a car or on a motorcycle through still air.

From experience with real winds and sound waves, it was known that the measured speed of sound in air depends on the wind speed and direction. For example, suppose the speed of sound in still air were 335 m/s. If there were a 25 m/s wind blowing in the direction in which the sound wave was propagating, then an observer would measure the sound speed to be $v = v_s + v_w = 335\ \text{m/s} + 25\ \text{m/s} = 360\ \text{m/s}$. Similarly, if the wind were blowing in the opposite direction, the observer would measure the sound speed to be $v = v_s - v_w = 335\ \text{m/s} - 25\ \text{m/s} = 310\ \text{m/s}$.

Thus, the motion of the air, or wind, changes the observed speed of sound. It was reasoned that the ether wind should do the same with the speed of light and thereby would prove that the ether existed.

This was the purpose of the famous Michelson and Morley experiment, which was performed in 1887 by two American scientists, A. A. Michelson and E. W. Morley.

Two equally powerful planes fly the same distance in a race. The plane flying perpendicular to the wind must steer slightly into the wind on both legs of the race, which slows the plane somewhat. The other plane, flying directly into the wind on the first leg of the race, is slowed much more, but the wind adds to the speed on the return leg. Even so, it can be shown by analysis or by an actual race that this plane always loses the race or takes a longer time to cover the distance. Your instructor can prove this to you.

If the atmospheric wind were replaced by the ether wind and the airplanes by light beams, the light beams would then arrive back at the finish line at different times, as the airplanes do. Michelson used an ingenious method to develop an instrument to measure the expected time difference. The method was based on the wave interference of light. The light beams interfering at a detector (finish line) produce a pattern of alternate bright and dark fringes. This pattern is noted for one position of the apparatus, which is call angle, say 90°. If this change in direction changes the travel times of the light beams, as it should in an ether wind, then there would be a shift in the interference pattern.

The interferometer was sensitive enough to detect a time difference resulting from adding or subtracting the Earth's orbital speed of 30 km/s to or from the 300 000 km/s speed of light, but nothing was observed! Measurements were made at different times (day and night, different seasons) and at different locations (America and Europe), but always with a null result. There was no fringe shift!

Where was that ether? Several explanations were suggested to make the null result of the Michelson-Morley experiment consistent with classical wave theory. Perhaps the ether in the immediate vicinity of the Earth was "dragged" along with it so the interferometer was at rest with respect to the ether. This would explain the lack of an interference fringe shift, since there would be no wind. But if this were the case, light waves coming to the Earth would also be dragged along, since they travel in the ether. As a result, we would always see the light from a distant star coming from the same direction, which is not the case.

Another popular explanation at the time was offered by G. F. Fitzgerald, an Irish physicist. He suggested that all objects contracted or shrank in the direction of their motion through the ether. This so-called Fitzgerald contraction would adequately explain the results of the Michelson – Morley experiment if the arm of the interferometer moving in the direction of the ether wind was shortened just enough so that no shift in the interference pattern would occur. It was a clever idea, but not without problems, since similar contractions weren't observed in other situations.

This was the unsettling state of affairs at the turn of the century that paved the way for a major reassessment of physical theories. In 1905 Albert Einstein published his special theory of relativity, which set forth the currently accepted explanation for the Michelson-Morley experiment. At the time, Einstein's theory was rejected by many scientists. It required some thinking that went against "classical common sense."

## 11.4 THE SPECIAL THEORY OF RELATIVITY

The special theory is based on two postulates:

1. The Principle of Relativity. All the laws of physics are the same for all observers moving at a constant velocity with respect to one another.

2. The speed of light in free space is the same ($c = 3 \times 10^8$ m/s) for all observers regardless of the motion of the source or the motion of the observer.

Let's take a look at each of these postulates to get a grasp of their meanings. The first postulate, the Principle of Relativity, has subtle implications. It is really an extension of Newtonian relativity for the laws of mechanics to all the laws of physics. The fact that physical laws or experimental results of these must be the same for all observers moving uniformly with respect to each other was recognized by Newton, as well as Galileo.

Another way of stating the first postulate is that the laws of nature are the same in a laboratory at rest as they are in any uniformly moving laboratory. Everything would appear the same. For example, if you were in a ship or an airplane moving with a constant velocity, the making and pouring of a cup of coffee or throwing something up and catching it would be exactly the same as when the ship or plane ere at rest. Or, looking at it another way, you can not tell, by any experiment, whether you are at rest or moving uniformly.

Suppose a person is watching uniformly moving automobiles. The velocities of the cars are shown referenced to the ground or the "stationary" observer. But if car A is taken as a reference system, car B is not moving and car C is moving with a speed of 10 km/h. As a matter of fact, with respect to car A, the "stationary" observer is moving with a speed of 40 km/h in the direction opposite to that of car C. Hence, the motion is relative.

But who is at rest? No one and everyone, in a sense. Anyone moving uniformly with respect to someone at "rest" is entitled to consider himself or herself to be at rest and the other person to be moving uniformly. A "state of rest", then, is one for which Newton's first law of motion (law of inertia) holds. We call this an inertial system—one at rest or in uniform motion.

What all this means is that there is no "absolute" reference frame with the unique property of being at rest with respect to everything else – like the ether. So, the ether is rejected in Einstein's theory.

### Summary of Key Terms

**Ether** A hypothetical medium for propagation of light waves.

**Michelson – Morley experiment** An experiment that was designed to detect the ether through velocity (vector) addition and interference. No effects were observed.

**Special theory of relativity** Einstein's theory of relativity that deals with non – accelerating or inertial systems.

**Principle of relativity** All the laws of physics are the same for all observers moving at a constant velocity with respect to one another.

**Constancy of light** The speed of light in free space is the same for all observers regardless of the motion of the source or the motion of the observer.

**Time dilation** The observation ($t$) of a clock in a moving system ($t_0$) running more slowly, according to the equation $t = \nu_0 t$.

**Length contraction** The observation ($L$) of a shortening of a length ($L_0$) in the direction of motion in a moving system, according to the equation $L = L/\nu_0$.

**Twin paradox** The paradox of a space – travelling twin returning to Earth younger than his Earth – bound twin, which is predicted by the general theory of relativity.

**Mass-energy conversion** The changing of mass into energy and vice versa, according to the equation $E = mc^2$. As a result, mass is considered to be a form of energy.

**General theory of relativity** Einstein's theory of relativity that deals with accelerated systems.

## Reading Materials:

### The Twin Paradox

Time dilation gives rise to another popular relativistic topic—the twin paradox. According to the result of the special theory, a clock in a moving system runs more slowly than a clock in an observer's system. For example, with a $\gamma = 4$, for every hour that elapses on an observer's clock, only one fourth of an hour ticks off on a clock in a moving system by his observations.

Similarly, observing one year of events in the moving system takes four years in the observer's time frame. Keeping in mind that we "age by the clock" (heartbeat and age measured by proper time), a good question quickly arises Does an observer in one system age more quickly than another in a relatively moving system?

The "twin paradox" states the problem in terms of a set of twins. Suppose one of the twins takes a high-speed space journey that takes 40 years according to the twin who stays on Earth. Would the Earth twin be older than the space twin I on return? With $r = 4$, the Earth twin would spend 40 years observing the 10 years (t.) that elapsed on the spaceship clock. If the twins were 25 years old at blast-off, then does the 35-year-old returning space I traveler find his twin getting ready for retirement at age 65? The answer is yes.

But what happens from the point of view of the space traveler? From the relative part of relativity, you know that the space twin looking back at the Earth "sees" his twin's clock running slowly. So from this point of view, the Earth twin ages more slowly, and he is the one who is younger. (Do not get the false idea that relativity allows one to go back in time. This is not the case.)

The difficulty with the analysis is that the space traveler does not remain in an inertial system. What happens when he is accelerating in starting and stopping? To investigate the paradox in detail, we should apply the general theory, which treats accelerating reference frames. However, by special applications of the special theory, it too predicts that the twin who has been accelerated will be younger than the one who stayed at home. You may be wondering how we know the "correct" answer, since this requires experimental testing.

The twin paradox has been tested. Not with real twins, of course, but with atomic-clock "twins". Four cesium atomic clocks were flown around the world in opposite directions on commercial air-craft. The clocks had been previously synchronized with stationary cesium-clock "twins" on Earth. Afterward, the moving clocks were found to be "out of sync" (showed a different time) in accord with the relativistic predictions. Ultra-accurate cesium clocks had to be used instead of common wristwatches because the time differences were on the order of billionths of a second. Even so, the flying clocks came back "younger."

In a more recently reported experiment (1985), rather than physically moving clocks, researchers used Earth-based clocks located in different countries. Pairs of these Earth stations simultaneously viewed signals from global positioning satellites, which, depending on the sequence observed, gives an east-west effect similar to that of the flying clocks. Time differences were noted as predicted by the theory of relativity. This research will help in synchronizing clocks around the world to subnanosecond accuracy.

Experiments with unstable particles accelerated to high speeds in particle accelerators also provide experimental testing that involves the twin paradox. These particles are stable for only a certain time, and it is observed that accelerated particles live longer, on the average, than those at rest or unaccelerated.

The twin paradox gives rise to such limericks as a precocious student quite bright could travel much faster than light. He departed one day in an Einsteinian way and arrived on the previous night.

The twin paradox. The theory of relativity predicts that identical twins (a) would age differently if one went on a space trip (b).

## Scientist: Albert Einstein

Albert Einstein (1879 ~ 1955) was one of the greatest figures in physics. Born on March 14, 1879, in Ulm, Germany, he received his early education in Germany. School instruction seemed to bore him, and he showed no particular intellectual promise. In 1896 at the age of 15, he entered the Swiss Federal Polytechnic school in Zurich to be trained as a teacher of physics and mathematics. In 1900 he received his diploma and acquired Swiss Patent Office in Berne, where his main duty was the preliminary examination of patent applications. This job provided a livelihood and left him with ample time to work on fundamental problems in physics. In 1905 he received a Ph.D. degree from the University of Zurich, and in the same year he published three papers of immense importance. Each contained a great discovery in physics. One dealt with quantum theory and an explanation of the photoelectric effect. This paper formed the basis for his receipt of the 1921 Nobel Prize in physics. Another addressed molecular motions and sizes and an analysis of Brownian motion. The third paper provided the special theory of relativity, which revolutionized modern ideas about space and time. Following this, Einstein had no difficulty in becoming a professor at various universities. In 1915, while at the University of Berlin, he published his paper on the general theory of relativity, which provided a new theory of gravitation that included Newton's theory as a special case. During the late 1920's the political situation in Germany deteriorated severely. In 1933, when Hitler and the Nazis came to power, Einstein immigrated to the United States and joined the Institute for Advanced for study in Princeton, New Jersey, where he settled permanently. He became a U.S. citizen in 1940. The rest of his life was spent working on a unified theory that would include both gravitation and electromagnetism. Near the beginning of World War II and following the discovery of nuclear fission, Einstein was asked to write a letter to President Roosevelt by other emigrant scientists, who along with him recognized the tremendous military potential of nuclear fission, particularly if Germany should develop it first. Einstein's famous letter was instrumental in starting the Manhattan Project and in developing the atomic bomb. After World War II, Einstein devoted much of his time to organizations advocating agreements to end the threat of nuclear war.

12

# The Nucleus and Radioactivity

## 12.1 THE ATOMIC NUCLEUS

Having looked at the orbiting "planet" electrons of the simplified planetary model of the atom, let's now look at the atomic "sun"—the nucleus of the atom. Another name for our model of the atom is the Rutherford – Bohr model. The Bohr contribution was discussed in the last chapter. The Rutherford part of the model is concerned with the central core or nucleus of the atom.

In the early 1900's the atom was thought to have a "plum-pudding" structure. This plum-pudding model viewed the electrons of the atom as being spread out like raisins in a sphere of positively charged "pudding." The pudding atom was thought to be about 10 ~ 8 cm across. But around 1910 there were experimental results that didn't agree with this model. The experiment was suggested by the British physicist Ernest Rutherford and was performed in his laboratory. Positively charged particles, called alpha particles, were directed toward a thin gold foil "target".

These alpha particles are thousands of times more massive than electrons, and the plum-pudding model predicted that they would be deflected only slightly as a result of collisions with the electrons. However, the results were completely different. Some of the alpha particles were scattered through large angles, even scattered backward. As Rutherford put it, this was almost as if you fired a 15-inch shell at a piece of tissue paper, and it came back and hit you.

In 1911 Rutherford offered an explanation of the alpha-particle scattering that gave a different view of the atom. If all the positive charges of an atom were concentrated in a central massive core or nucleus, then an alpha particle coming close to this region would experience a large deflecting force and would even be back—scattered in head—on collisions. Theoretical calculations showed that the theory fit the data, and the atomic nucleus was "born". For his efforts, Rutherford was given a title (he became Lord Rutherford) and received a Nobel Prize.

The alpha scattering is an example of a "black box" experiment that is very common in nuclear physics. In general, you know what goes in, observe what comes out, and infer what happened in between. The atomic nucleus was Rutherford's "black box". The 180° back—scattering gives an upper limit on the nuclear size or radius. (The nuclear size must be less than a volume with a radius equal to the distance of closest approach.) This is on the order of 10 ~ 14 m for a typical atom. The atomic electrons in the Rutherford – Bohr model are much farther out, on the order of 10 ~ 10 m.

Because of the small size of the atom, most people do not realize the extent of the "void of space" within the atom. The relative dimensions of the atom's structure have been likened to a large major league football stadium with the nucleus being a small marble (on the 50-yd line). If the nucleus of a typical atom were expanded to the size of the Sun, then on a proportionate basis, the "inner planets" (Mercury and Venus) would be well beyond the outer reaches of our solar system—many, many times beyond.

## 12.2 NUCLEAR NOTATION AND ISOTOPES

Recall that the number of nuclear protons in an atom determines what kind of atom it is. The nucleus of a hydrogen atom is a single proton, helium atoms have two protons, lithium atoms have three protons, and so on. However, it was found that all of the nuclear mass could not be accounted for by protons. Rutherford suggested in 1920 that there may be another electrically neutral particle in the nucleus, which he called a neutron. The existence of the neutron was confirmed experimentally in 1932. We now know that all nuclei, with the exception of the common hydrogen nucleus (a proton), contain neutrons, which are electrically neutral particles with a mass about the same as that of a proton.

Protons and neutrons are about 2 000 times more massive than electrons, so the vast majority more than 99.9 percent of the atomic mass lies in the nucleus. Nuclei are generally thought of as being spherical, but they may deviate from this shape and resemble a watermelon or a doorknob. Nuclear protons and neutrons are collectively referred to as nucleons, but we speak specifically of the proton number (sometimes called the atomic number) and the neutron number of a nucleus which of course are just the numbers of protons and neutrons in a particular type of nucleus.

To designate a particular nucleus or nuclear species, which is called a nuclide, a special notation, for a general case and a carbon nuclide, is used. Notice that the proton number is on the lower left of the chemical symbol of the element or atom. The proton number is also commonly called the atomic number and is designated by the letter Z. The fact that a nucleus has six protons makes it a carbon nucleus. The neutron number is on he lower right.

Notice that if you add the number of protons and the number of neutrons (6 + 6), you get the number(12) to the upper left of the symbol.

This is called the mass number $(p + n)$. It is customary to leave off the neutron number and simply write $^{12}_{6}C$. This still gives you the same information. In referring to this nucleus, we say that it is a carbon – 12 nucleus.

If you examined a bunch of carbon nuclei (all with six protons), you would occasionally find a nucleus with more than six neutrons—sometimes seven neutrons and, on rarer occasions, eight neutrons. Nuclei or nucleons of the same element having different numbers of neutrons are called isotopes. A group of isotopes is somewhat like a family—they are all Joneses or Smiths (for example, hydrogen or carbon), but the individual family members are different. In the nuclear case, the "family" members differ by the number of neutrons they contain. Some elements have large nuclear isotope "families" with a dozen or more members.

Only the isotopes of hydrogen are given specific names. $^{1}_{1}H$, the most common isotope, has just a nuclear proton and is referred to as ordinary hydrogen or simply hydrogen. The other isotopes, $^{2}_{1}H$ and $^{3}_{1}H$, are called deuterons and tritones, respectively, or deuterium and tritium in atomic form. When the more massive deuterium atom replaces the common hydrogen atom in $H_2O$, we have what is called "heavy water". For every 6 500 or so atoms of ordinary (light) hydrogen in water, there is one atom of deuterium. The oceans contain millions of tons of deuterium. Tritium is radioactive. You will learn more about both these isotopes later in the discussion on nuclear energy. Other isotopes are referred to by their element—mass number designation, for example, carbon – 12, carbon – 13, and carbon – 14.

## 12.3 THE NUCLEAR FORCE

Within the small confines of the nucleus, the nucleons are clustered together, with each nucleon presumably taking up about the same amount of space. The nucleons are on the order of 10 ~ 15 m apart. This means that there are large repulsive electrical forces between the positively charged protons. What then holds the nucleus together? Since the nucleons are mass particles, the ever-present attractive gravitational force is there. However, calculations show that the gravitational force between two nucleons is a factor of $10^{40}$ smaller than the electrical force between two protons. On a relative basis, with a factor of $10^{40}$, the gravitational force is so small that it is negligible, so that's out.

Obviously, there must be some other strongly attractive force acting in the nucleus that overcomes the repulsive electrical force. We call this attractive force acting between nucleons the nuclear force, or the strong interaction. The nuclear force is a fundamental force like the electrical and gravitational forces, but it is more complicated and not completely understood. From experiments, we believe that this force acts between any pair of nucleons-rotor-proton, neutron-neutron, and proton-neutron.

The strong nuclear force is a short-range force. That is, it falls off very quickly with nucleon separation distance. Recall that the lectrical force falls off as the inverse square of the distance $1/r^2$. The strong nuclear interaction weakens much more quickly. For example, the repulsive electrical force between two protons on opposite sides of a sizable nucleus may be appreciable, but the attractive nuclear force is very small. An electrical force also exists between the nuclear protons and the orbiting atomic electrons (centripetal force), but the nuclear force does not extend outside the nucleus.

If the attractive nuclear and repulsive electrical forces acted only between protons, then there would be only small nuclei or atoms because of the short range of the nuclear force. However, there are very massive stable nuclei, for example, the lead isotope $^{208}_{82}Pb$, with 82 protons and 126 neutrons. The key is the number of neutrons. Remember that the nuclear force acts between all nucleons, so the neutrons act as a nuclear "glue" to hold the nucleus together. As the number of protons increases for massive stable nuclei, the number of neutrons increases too, so that the short-range nuclear forces are greater than the long – range repulsive electrical forces. Neutrons provide attractive nuclear forces between both protons and other neutrons.

For lighter or less massive nuclides up to about calcium (proton number of 20), the stable nuclei have an equal or an approximately equal number of protons and neutrons. But for heavier nuclides, there are more neutrons than protons.

If you look at the magnitudes of the attractive nuclear forces between nucleons and the attractive electrical forces between nuclear protons and orbiting atomic electrons, the latter are about one billionth as great. This gives a hint about why the nuclear energy of atoms is so much greater than chemical energy.

In a chemical reaction, such as the burning or combustion of gasoline, there is a rearrangement of the electrons in the atoms or molecules. The nuclei of the atoms do not change. In a nuclear reaction, however, there is a "rearrangement" or change of nuclear particles. This involves strong forces and the release of large amounts of energy, as will be learned in the next chapter in the discussion on nuclear energy.

## 12.4 RADIOACTIVITY

For some reason, the previous argument about neutron "glue" doesn't apply, and unstable nuclei spontaneously "decay" with the emission of energetic particles. Such nuclei are said to be radioactive. The "radio – " part refers to the emitted radiation, which in the modem context may be a particle or a wave. So a radioactive nuclide or isotope is "active" in emitting radiation. There are no stable nuclei with proton numbers greater than 83.

Of the nearly 1 200 known unstable nuclides, only a small number occur naturally. The other radioactive nuclei are made artificially. The unstable nuclides found in nature decay with the emission of alpha particles or beta particles, which are sometimes accompanied by gamma rays. Using a magnetic field, which deflects electrically charged particles, it was found that alpha "rays" were positively charged particles and beta "rays" were negatively charged particles. Because they are deflected more, beta particles must be less massive than alpha particles. Gamma rays were not deflected at all, so they must be uncharged. Actually, a gamma ray or "particle" is a photon or quantum of energy.

Radioactivity was discovered accidentally by the French physicist Henri Becquerel in 1896. The circumstances were not unlike Roentgen's discovery of X-rays. Becquerel noticed that a hotographic plate in a light-tight wrapper that he had left in a drawer with a uranium compound showed signs of exposure when developed. Evidently, some type of radiation or "rays" from the uranium was able to penetrate the wrapper and expose the film.

A couple of years later, the husband-and-wife team of Pierre and Marie Curie announced the discovery of two new radioactive elements, radium and polonium, which they had isolated from uranium pitchblende ore. They had painstakingly isolated 10 mg of radium and a smaller amount of polonium from 8 tons of ore! The Curies and Becquerel shared the 1903 Nobel Prize in physics for their work in radioactivity. Madame Curie, as she is commonly known, also received the Nobel Prize in chemistry in 1911 for her contributions in chemically isolating radioactive materials. Carrying on the family tradition, the Curies' daughter, Irene Joliot Curie, and her husband Frederic Joliot won the 1935 Nobel Prize in chemistry.

Let's take a closer look at the ABCs of radioactivity: alpha($\alpha$) beta ($\beta$), and gamma ($\gamma$) decays and their respective particles.

## Summary of Key Terms

**Atomic nucleus** The central core of the atom in which the protons and neutrons of an atom are located.

**Nucleon** A nuclear proton or neutron.

**Nuclide** A particular nucleus or nuclear species.

**Proton number** The number of protons in a nucleus, which defines its atoms as being a particular element.

**Mass number** The sum of the protons and neutrons in a nucleus. energy of a particle into visible light.

**Isotopes** Nuclei or nuclides of the same element with different numbers of neutrons.

**Nuclear force** The strong attractive interaction that acts between nucleons.

**Radioactivity** The spontaneous decay of certain isotopes with the emission of energetic particles.

**Alpha particle** A particle consisting of two protons and two neutrons, which is the same as a helium nucleus.

**Beta particle** An electron.

**Gamma particle** A quantum or photon of energy.

**Half - life** The time it takes for one half of the nuclei of a sample of a given radioactive isotope to decay.

**Geiger counter** A common radiation detector based on the ionizing nature of radiation.

**Dead time** The time required for a detector to recover for another detection or count.

**Scintillation counter** A radiation detector based on the ability of a phosphor material to convert the fundamental interactions in a unified field theory.

**Elementary particles** The fundamental building blocks of nature, i.e., the basic or "elementary" particles that make up matter and account for the interactions thereof.

**Exchange particles** The elementary particles responsible for fundamental forces or interactions.

**These are as follows** For the strong force, the gluon; for the weak force, the *W* and *Z* particles; for the electromagnetic force, the photon; and for gravity, the graviton.

**Hadron** Any panicle that interacts by the strong force.

**Quarks** Subparticles with fractional electronic charges thought to make up hadrons.

**Electroweak force** The single force believed to incorporate both the electromagnetic and weak forces.

**Grand unified theory (GUT)** A theory that combines the strong and electroweak forces into a single force.

**Superforce** A single force that would describe all.

## Reading Materials:
## Medical Applications of nuclear radiation

In medicine, nuclear radiation has both advantages and disadvantages. It can be used beneficially in the diagnosis and treatment of some diseases but can also be potentially harmful if it is not properly handled and administered. Nuclear radiation and X-rays can penetrate human tissue without pain or any other sensation. However, early investigators quickly learned that large doses or repeated small doses led to red skin, lesions, and other conditions.

The chief hazard of radiation is damage to living cells, primarily due to ionization. Several effects can occur. Ions, particularly complex ions or radicals, produced by ionizing radiation may be highly reactive (for example, a hydroxyl ion $OH^-$ from water). These interfere with the normal chemical operations of the cell. Also, radiation can alter the structure of a cell so that it cannot perform its normal functions. The cell may even die. This might not be serious, since similar cells would reproduce to make more cells. But if enough cells are damaged with large radiation doses, cell reproduction might not be fast enough, and the irradiated tissue could eventually die.

In other instances, there may be damage to a chromosome in the cell nucleus cell(genetic damage). Again, the cell might die, but it could instead live with damaged genetic code for cell reproduction. Damaged cells might reproduce, but the new cells might be different or defective. Also, such reproduction often occurs at an uncontrolled rate. This rapid, unregulated production of abnormal cells is called cancer.

Cancer cells may grow slowly in number with little effect on the surrounding normal cells, forming a benign tumor. They also may grow at the expense of the surrounding cells, producing a malignant tumor. Skin cancer and leukemia (cancer of the blood) are probable results of excessive radiation exposure Leukemia is characterized by an abnormal increase in the number of white blood cells (leukocytes). The human cells most susceptible to radiation damage are those of the reproductive organs, bone marrow, and lymph nodes.

Radiation can cause cancer, but it can also be used to treat cancer. In this case, cell damage by radiation is used to control the growth of cancer cells. The radiation may be gamma rays from a radioactive $^{60}Co$ source or X-rays from a machine. Other particles produced in particle accelerators are also used to treat cancer. The electrical charge and energy of the particles determine the penetrating power of the radiation. X-rays and gamma rays are deeply penetrating, but generally, electrons (beta particles) penetrate only a few millimeters into biological tissue, and alpha particles Penetrate only a fraction of a millimeter.

An important consideration in radiation therapy and radiation safety is the amount or dose, of radiation. Several quantities are used to describe this in terms of exposure, absorbed dose, and equivalent dose. The earliest unit of dosage, the roentgen (R), was based on exposure and was denned in terms of the ionization produced in air. (One roentgen is the quantity of X-rays or gamma rays required to produce an ionization charge of $2.58 \times 10^{-4}$ C/kg of air.) The roentgen has been largely replaced by the rad (radiation absorbed 'dose) whiten is an absorbed dose unit. One rad is an absorbed dose of radiation of 10 – 2 J of energy per kilogram in any absorbing material. The SI unit for absorbed dose is the gray (Gy):

$$1 \text{ Gy} = 1 \text{ J/kg}$$

These physical units give the energy absorbed per mass, but it is helpful to have some means for measuring the biological damage produced by radiation, since equal doses of different types of radiation produce different effects. For example a relatively massive alpha particle with a charge of 2 + moves through tissue rather slowly with a great deal of Coulomb interaction. The ionizing collisions thus occur close together along a short penetration path, and more localized damage is done than by a fast-moving electron or gamma ray.

This effect or effective dose, is measured in terms of the rem unit (roentgen or rad equivalent man). The different degrees of effectiveness of different particles are characterized by the relative biological effectiveness (RBE). The RBE is denned in terms of the number of rads of X-rays or gamma rays that produces the same biological damage as one rad of a given radiation.

The effective dose is then given by the product of the dose in rads and the appropriate RBE: effective dose (in rem) = dose (in rad) × RBE Thus, one rem of any type of radiation does approximately the same amount of biological damage.

For example, a 20 – rem effective dose of alpha particles does the same damage as a 20 – rem dose of X-rays, but 20 rad of X-rays are needed compared to 1 rad of alpha particles.

The SI unit of absorbed dose is the gray, and the effective dose with this unit is called the sievert (Sv):

effective dose (in Sv) = dose (in Gy) × RBE

Since 1 Gy = 100 rad, it follows that 1 Sv = 100 rem.

It is difficult to set a maximum permissible radiation dosage, but the general standard for humans is an average dose of 5 rem/year after the age of 18 with no more than 3 rem in any 3-month period. In the United States, the normal average annual dose per capita is about 200 mrem (millirem). About 125 mrem comes from the natural background of cosmic rays and naturally occurring radio-active isotopes (in the soil, building materials, and so on). The remainder is chiefly from diagnostic medical applications, mostly X-rays, and from miscellaneous sources such as television tubes.

Radioactive isotopes offer an important technique for diagnostic procedures. For example, a radioactive isotope behaves chemically like a stable isotope of the element, even though it is radioactive. That is, the radioactivity is independent of the chemical compound in which the isotope resides, and the atom of an isotope can participate in chemical reactions without affecting the radioactivity. This makes it possible to label or tag molecules with radio isotopes, which can then be used as tracers.

By using tracers, many body functions can be studied simply by monitoring the location and activity of the labeled molecules as they are absorbed in the body processes. For example, the activity of the thyroid gland in hormone production can be determined by monitoring its iodine uptake using radioactive 131. Similarly, radioactive solutions of iodine and gold are quickly absorbed by the liver. This can be done safely because the isotopes have short half-lives. One of the most commonly used diagnostic tracers is technetium – 99 (Tc). It has a convenient half-life of 6 hours and combines with a large variety of compounds. Detectors outside the body can scan over a region and record the activity so that a complete activity image may be reconstructed.

It is possible to image gamma ray activity in a single plane or "slice" through the body. A gamma detector is moved around the patient to measure the emission intensity from many angles. A complete image can then be constructed by using computer-assisted tomography (from the Greek tomo-, meaning "slice") as in X-ray CAT scans. This is referred to as single-photon emission tomography (SPET). Another technique, positron emission tomography (PET), uses tracers that are positron emitters. These include C and O. When a positron is emitted, it is quickly annihilated, and two gamma rays are produced. Recall that the photons have equal energies and fly off in nearly opposite directions so as to conserve momentum. The detection of the gamma rays is by a ring of detectors surrounding the patient.

Another medical tool is nuclear magnetic resonance (NMR). A compass needle or magnet tends to line up in a magnetic field. The orbital motions of atomic electrons give rise to current loops resulting in magnetic moments to which the intrinsic electron spin also contributes. When atoms are placed in a magnetic field, the atomic energy levels are split into several closely spaced levels.

Many nuclei exhibit similar properties and have magnetic moments. For example, the simple hydrogen nucleus can have only two values of spin, similar to its electron. When a hydrogen atom is placed in a magnetic field, the energy levels split, and we have a "spin-up" state and a "spin-down" state, which correspond to the direction of the spin being parallel and antiparallel, respectively, to the magnetic field. It turns out that the spin-up state has the lower energy.

If varying radio-frequency (r - f) radiation is applied to a sample containing hydrogen nuclei, as all tissues do, when the frequency is in resonance or equal to that of the energy difference of the spin levels ($E = hf$), photons are absorbed. This is the basis of the name "nuclear magnetic resonance." The spin directions are reversed in the process. By measuring the emitted radiation when the nucleus returns to the lower state and using tomography techniques, computers construct an image based on the directions and intensities of the reradiated photons.

Although a variety of atoms or nuclei exhibit nuclear magnetic resonance, most medical work is done with hydrogen atoms because of the varying water content of tissue. For example, muscle tissue has more water content than fatty tissue, so there is a distinct contrast in radiation intensity. Fatty deposits in blood vessels are distinct from the tissue of the vessel walls. (The water-rich blood is not seen because it has moved to a new tomography plane by the time it reradiates.) A tumor with a water content different from that of the surrounding tissue would show up in an NMR image.

## Scientist: Marie Curie

Marie Curie (1867 ~ 1934) was born in Poland and studied in France. There she met and married Pierre Curie (1859 ~ 1906), who was a physicist well known for his work on crystals and magnetism. In 1903, Madame Curie (as she is commonly known) and Pierre shared the Nobel Prize in physics with Henri Becquerel for their work on radioactivity.

Marie Curie was also awarded the Nobel Prize in chemistry in 1911 for the discovery of radium and the study of its properties.